SpringerBriefs in Latin American Studies

Series Editors

Jorge Rabassa, Lab Geomorfología y Cuaternario, CADIC-CONICET, Ushuaia,
Tierra de Fuego, Argentina

Eustógio Wanderley Correia Dantas, Departamento de Geografia, Centro de
Ciências, Universidade Federal do Ceará, Fortaleza, Ceará, Brazil

Andrew Sluyter, Conference of Latin Americanist Geographers, Louisiana State
University, Baton Rouge, LA, USA

More information about this series at http://www.springer.com/series/14332

Rubson Maia · Francisco Bezerra

Structural Geomorphology in Northeastern Brazil

 Springer

Rubson Maia
Departamento de Geografia
Universidade Federal do Ceará
Fortaleza, Ceará, Brazil

Francisco Bezerra
Universidade Federal do Rio
Grande do Norte
Parnamirim, Rio Grande do Norte, Brazil

ISSN 2366-763X   ISSN 2366-7648   (electronic)
SpringerBriefs in Latin American Studies
ISBN 978-3-030-13310-8   ISBN 978-3-030-13311-5   (eBook)
https://doi.org/10.1007/978-3-030-13311-5

This Springer imprint is published by the registered company Springer Nature Switzerland AG
The registered company address is: Gewerbestrasse 11, 6330 Cham, Switzerland

Introduction

In Northeastern Brazil, the relief documents important episodes in the morpho-tectonic evolution. Organized around paleosurfaces, the region presents diverse geomorphological compartments derived from tectonic events, such as the Brasiliano cycle and the breakup between South America and Africa. Both events generated various structural morphologies. The crystalline massifs that are aligned according to different shear zones are worthy of note. Other forms are associated with structural lineaments that control drainage and dissection. The third type are the Mesozoic Basins affected by uplift. This whole group comprises a complex morphostructural system, which began to be interpreted from the 1960s as composed of successive levels of terraced paleosurfaces. However, the region has a great number of Cenozoic deformation structures, especially in sedimentary areas, sometimes influencing the sediment deposition and erosion.

From this perspective this book discusses the models of geomorphological evolution of Northeastern Brazil, analyzing their features and limitations with regard to the suitability of the concepts associated with the Cenozoic tectonic and the geochronology of the geological units. It also deals with aspects related to relief, starting from individualized studies developed in areas with proven Cenozoic tectonic activity, such as the Potiguar Basin in the states of Ceará and Rio Grande do Norte. The analytical focus involves the relationship between tectonics and relief on different scales and geological times, be it of detail as in the case of the valleys and karst, or regional as in the case of basins and massifs.

Contents

1   **The Paradigm of Stable Intraplate Regions and Neotectonics**
    in Northeastern Brazil   1
    1.1   The Stable Crust Paradigm   1
    1.2   The South American Plate   2
    1.3   Seismicity in Northeastern Brazil   4
    1.4   The Causes of Intraplate Seismicity and Examples
         of Seismogenic Faults   6
    1.5   Present-Day Stress Field   8
         1.5.1   Focal Mechanisms   8
         1.5.2   Borehole Breakouts   9
         1.5.3   Image Logs 10
         1.5.4   Fault-Slip Data in Quaternary and Miocene
               Sedimentary Units 11
    References 11

2   **Understanding the Geological Setting of Northeastern Brazil** 15
    2.1   Introduction 16
    2.2   The Borborema Province 16
    2.3   The Cretaceous Basins 18
    2.4   The Neogene–Quaternary Sedimentary Covers 19
    2.5   Introduction to Neotectonics in Northeastern Brazil 20
    2.6   Paleoseismicity in Northeastern Brazil 21
    2.7   Miocene to Quaternary Tectonics and Sea-Level Changes 23
    References 26

3   **The Geomorphology of the Northeast: Classical and Current**
    Perspectives 31
    3.1   Classic Models of Geomorphological Evolution 31
    3.2   Genetic Aspects of the Relief of Northeastern Brazil:
         Classic Concepts 33

3.3   Synthesis of the Weak Points of the Paleosurface Model   35
3.4   Current Concepts About the Geomorphology of Northeastern
      Brazil.   35
References   39

4 The Erosion and Exhumation of Massifs in Precambrian
   Shear Zones   41
   4.1   Influence of Fault Reactivations on the Relief Evolution
         in the Borborema Province   44
   4.2   Structural Control of the Relief and Drainage: Examples
         from the Borborema Province   45
   4.3   Structural Controls   50
   4.4   Exhumation of the Brasiliana Shear Zones   51
   4.5   Differential Erosion and Topographical Inversion   53
   4.6   Granitic Inselbergs   55
   References   67

5 The Morphostructural Evolution of Cretaceous Basins.   71
   5.1   Neotectonic Stress Field and Basins Inversion   72
   5.2   Geological and Geomorphological Characterization
         of the Potiguar Basin   73
   5.3   Extensional Basins and Compressional Tectonics   75
   5.4   Sedimentary Tectonics and Basin Inversion   79
   References   83

6 Karstic Geomorphology   85
   6.1   Geological and Environmental Characterization   88
   6.2   Post-rift Tectonic and Faults in the Potiguar Basin   90
   6.3   Karstic Morphology   91
   References   99

7 Neotectonics and River Valleys   101
   7.1   The Geomorphology of the Northeast: Genetic Aspects   103
   7.2   Tectonics and Fluvial Systems   105
   7.3   The Morphotectonic Evolution of the Brazilian Northeast   106
   References   111
   Bibliography   113

Chapter 1
The Paradigm of Stable Intraplate Regions and Neotectonics in Northeastern Brazil

Abstract Active faults in stable continental regions are less frequent than those in plate boundaries but cannot be ignored. These faults can generate earthquakes in intraplate settings that cause widespread damage in regions where the population is ill-prepared to cope with them. Intraplate earthquakes are due to many causes such as reactivation of ancient rifted crust, major terrane boundaries, density and rigidity contrast, heat flow, and fault intersection. The present-day stress field in the South American plate roughly follows the trajectories of the absolute plate, i.e., is mainly E-W-oriented. In Northeastern Brazil, the stress regime is mainly strike-slip and the maximum horizontal stresses follow the equatorial coastline. Northeastern Brazil is one of the most seismically active parts of the South American stable continental region and presents earthquakes up to magnitude Mb = 5.2. Seismicity is concentrated in the upper crust down to a depth of 12 km and illuminates active faults up to 40 km long. Active faults either reactivate shear zones or regional foliation and quartz veins or cut across the preexisting tectonic fabric.

Keywords Faults · Neotectonic · Northeastern · Brazil

1.1 The Stable Crust Paradigm

Several names have been used to describe intraplate areas. These names include passive margins, stable continental margins, stable continental crusts, stable regions, and cratons. Most of these names are relative definitions that compare intraplate areas with active tectonic plate boundaries. However, part of the scientific community still sees intraplate areas as devoid of tectonic activity.

Intraplate areas occur away from plate boundaries and have not experienced any major orogenic cycle since the Cretaceous or any major rifting episode since the Paleogene (Schulte and Mooney 2005). Intraplate areas exhibit a long period of seismic recurrence and poor surface expression of earthquakes. Active faults are rare in intraplate areas when compared with those that occur at plate margins. The period of recurrence of active faults in intraplate areas can range from 10,000 to

100,000 years or more (Crone et al. 1997). For example, only ten great earthquakes produced surface ruptures in intraplate regions in the twentieth century (Crone et al. 2003).

However, despite the low seismicity, even moderate seismic activity in intraplate settings could cause widespread damage because the buildings and the populations are less prepared to cope with it (Hanks and Johnston 1992). Increasingly, investigations have indicated that although the tectonic activity in intraplate areas is not comparable with that observed at plate boundaries, this intraplate activity could not be ignored. An increasing number of studies have supported that intraplate areas are more active than previously supposed (e.g., Bezerra and Vita-Finzi 2000; Japsen et al. 2012; Gurgel et al. 2013). These studies have pointed out the occurrence of an important post-rift activity since the last major rifting phase in the early Cretaceous.

The need to improve knowledge of the tectonic activity and landscape evolution in intraplate areas has increased significantly. Several scientific gaps still remain and include the following: (1) understand how fault propagation influence sediment deposition and erosion of landforms, as well as the creation of new relief; (2) The sedimentation and morphology have been attributed to factors such as tectonics and climate (Bezerra et al. 2008). However, the contribution of individual factors such as tectonics and climate in shaping the morphology and sediment supply has been difficult to isolate (Bezerra et al. 2007); (3) There is an overall acceptance that some kind of tectonic activity could have occurred during sediment deposition, but much has to be investigated about how sediments have been deposited and deformed (Dentith and Featherstone 2003).

1.2   The South American Plate

The relatively stable interior of the South American plate is limited by the Andes to the west and by MAR to the east. The plate interior is composed of a continental crust and an oceanic crust (Fig. 1.1). The South American plate is currently moving westward away from the middle ocean ridge and against the Nazca Plate and is subjected to compression from both east and west. The velocity of spreading at the ocean ridge is ~34 mm/year. The Nazca Plate, however, moves toward the South American Plate at a velocity ~84 mm/year (DeMets et al. 1990).

The target area of this book is the Atlantic margin of the South America plate, whose geometry and limits are well known. The eastern and western edges of this plate are better defined than its northern and southern boundaries (Minster and Jordan 1978; DeMets et al. 1990; Meijer and Wortel 1992). The eastern boundary is marked by shallow seismicity along the mid-ocean ridge in the South Atlantic, where spreading rates ranges from 32 to 38 mm/year at 5 S to 40 S (Gordon and Jurdy 1986). Along the western boundary, the Nazca plate subducts at a rate of approximately 84–90 mm/year along a trench ~6000 km long (Dewey and Lamb 1992). In the northern margin of the plate, the boundary is broadly defined by the Caribbean and North American plates (DeMetes et al. 1990). The southern boundaries occur along

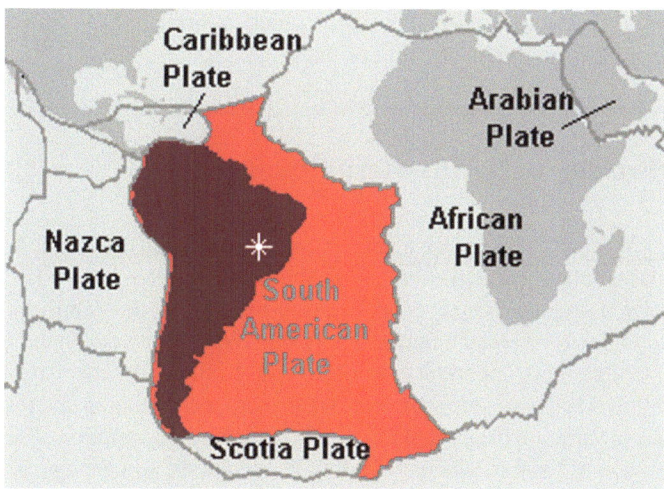

Fig. 1.1 The South American plate with the continental area (con) and oceanic area (oce). The main boundaries of the plate are in the mid-ocean ridge (mar) with the African plate and the subduction zone (sub) with the Nazca plate

the South Scotia and North ridges, where the South American and Antartic plates converge at a rate of 21 ± 02 mm/year (Minster and Jordan 1978).

Three main tectonic units occur within the South American plate: the Andes Mesozoic-Cenozoic orogenic belt, the continental cratonic region (Precambrian shields and Paleozoic to Cenozoic platforms), and an oceanic crust in the ocean (Almeida et al. 2000). Each main tectonic unit has a distinct topography related to its evolution and age.

The Precambrian shields and Paleozoic to Cenozoic platforms may be subdivided into a number of geological provinces and are the core of the present work. The Precambrian shield includes the Atlantic, Central Brazil, and Guyana shields, whereas the Paleozoic to Cenozoic platforms encompass most of the sedimentary basins on the continent (Almeida et al. 1981; Brito Neves et al. 2014). The study area is located in the Atlantic shield, which is a crystalline region composed of Archean and Proterozoic terrains. These regions encompass ancient mobile belts formed or deformed during the Brasiliano Orogeny (740–540 Ma,) and cratons, which remained stable during this orogeny (Brito Neves et al. 2014).

The South American Platform comprises major inland Paleozoic to Mesozoic Basins, and Mesozoic Basins along the continental margin (Almeida et al. 2000). The coastal provinces, where the studies presented in this book were carried out, overlies more than 8000 km of the Atlantic margin. This margin was developed during the Africa–South America breakup. The Northern Brazilian equatorial margin opened as a transform margin, which resulted in shear-dominant basins, whereas the East Brazilian margin formed by crustal extension (Mascle 1976). This book presents studies located within the Borborema Province in the Atlantic shield and the mostly

Mesozoic to Cenozoic Coastal Province, which is bordered by an oceanic crust formed along the South Atlantic mid-ocean ridge.

1.3   Seismicity in Northeastern Brazil

Intraplate earthquakes take place at unexpected sites previously regarded as aseismic such as places where no active fault or seismic activity were known (Bezerra et al. 2006). These unexpected occurrences contribute to the fact that intraplate seismicity is not well understood when compared with seismicity at plate boundaries (Bezerra et al. 2007). The most important scientific gaps of intraplate seismicity include the determination if active seismogenic faults reactivate preexisting structures or are new structures, the variation of stress regimes represented by the variation of several stress indicators such as focal mechanisms, borehole breakouts, and geological indicators, the recurrence rate of seismic events, and the determination of seismic zones where earthquakes are likely to occur. We will discuss these gaps in this section.

The earthquake historical record in intraplate South America spans less than 250 years (Branner 1912, 1920; Assumpção et al. 2014). The maximum event of this intraplate area was a 6.2 Mb earthquake in the west part of Brazil (Barros et al. 2009), but no evidence for surface faulting in this epicentral area was carried out. About half of the events greater than 5.0 Mb occurred along the Atlantic margin (Berrocal et al. 1984). Seismicity in intraplate Brazil, which corresponds to most of the Precambrian shields and sedimentary platforms of South America, has events with magnitudes 5 ou more with a return period of 4 years. Earthquakes usually happen in seismogenic zones, with large "aseismic" zones in between, but no single model has been able to explain the existence of these seismogenic zones. Seismicity is concentrated in Neoproterozoic belts in areas of thin crust or near craton edges. In addition, earthquakes are three times more likely to occur near mapped neotectonic faults and are concentrated along the Atlantic coast (Assumpção et al. 2014).

Northeastern Brazil is one of the most seismically active areas in intraplate South America (Fig. 1.2). The region is the source of a large part of the historical and recent seismicity. Both the seismic historical accounts and the instrumental seismic catalog indicate that the region is one of the most seismically active parts of intraplate South America. It is characterized by long-lasting earthquake sequences (e.g., Ferreira et al. 1997, 1998, 2008). In addition, the peak ground acceleration, the maximum acceleration that occurred on the ground during an earthquake varies from 0.2 to 2.4 in this region (Shedlock and Tanner 1999).

The historical seismic record of the region is short, patchy and is biased toward populated areas along the coast. The record dates back to the seventeenth century. The first historical account of an earthquake occurred in 1666 in the Recôncavo area, near the city of Salvador (Fig. 1.1) and was related to a possible tsunami that reached the shores of the area (Branner 1912, 1920) (Fig. 1.1). Along the Equatorial margin, the first event occurred in 1808 close to the city of Assu (Fig. 1.1). This event has an estimated 4.8 magnitude and Modified Mercalli intensity MM $\geq$ VI (Ferreira and

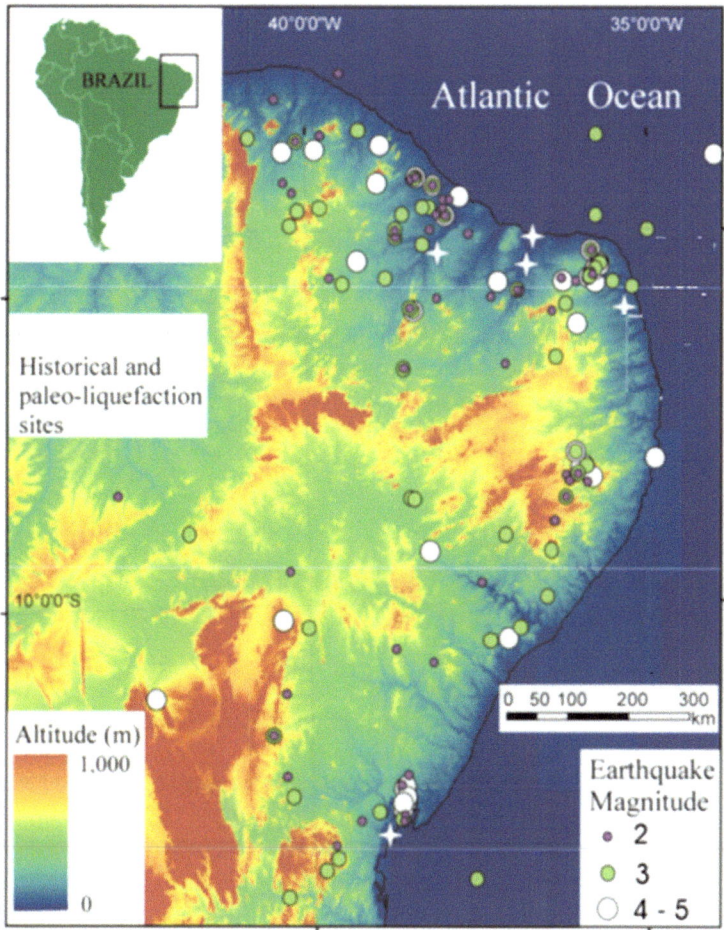

Fig. 1.2 Seismicity in the eastern part of Northeastern Brazil. Epicenters are from the selected telemetric-network dataset (Laboratory of Seismology at the Federal University of Rio Grande do Norte) (modified from Bezerra et al. 2007)

Assumpção 1983; Berrocal et al. 1984). It caused damage to the building, including total collapse and liquefaction. However, no surface rupturing event has ever been reported in the region until the present (Bezerra et al. 2006, 2011).

The seismicity has been instrumentally recorded since the 1960s. Earthquakes sequences in the region can last more than a decade (Ferreira et al. 1998; Takeya et al. 1989; do Nascimento et al. 2004). The turning point of the investigation of seismological studies in the region was the 1986–1989 seismicity of the João Câmara area (see details below). After this seismicity, local seismograph networks were progressively installed to monitor seismic sequences (Ferreira et al. 1998). The monitoring of seismic sequence is usually carried out after the local population feels a series of events and it usually monitors aftershock sequences (Ferreira et al. 1998, 2008; do Nascimento et al. 2004).

The instrumental monitoring of the seismicity indicates that a few features are common. The depth of the events ranges from 1 to 12 km. Events are usually moderate to small earthquakes. The major event to strike the region was a 5.2 Mb earthquake in 1980 at the western part of the Potiguar Basin. Only a few earthquakes reached magnitudes closer or larger than 5.0 Mb in the past 200 years, but the long duration of the seismicity has allowed the study of the activity of many faults in the region (Ferreira et al. 1998; Bezerra et al. 2011).

Most of the faults defined seismically occur onshore and from 20 to 300 km from the coast. Both the historical and instrumental seismicity indicate that the region exhibits three main seismic zones: The crystalline border of the Potiguar Basin, the east part of the Pernambuco lineament, and the west part of the Borborema Province along the Equatorial margin and close to the Parnaíba Basin (Ferreira et al. 1998, 2008; Bezerra et al. 2011).

1.4   The Causes of Intraplate Seismicity and Examples of Seismogenic Faults

The intraplate earthquakes have been associated with the brittle failure of preexisting rift faults (Johnston 1996; Rao et al. 2002), fault intersections (Talwani 1988), major terrain boundaries (Dendith and Featherstone 2003), ductile shear zones (Zoback 1983), boundaries of mafic intrusions (Campbell 1978), transform faults in the ocean (Sykes 1978), and rigidity contrasts (Stevenson et al. 2006). Other studies draw attention that usually intraplate seismicity has a weak correlation with surface features and does not favor specific structures, but indicate that seismicity occurs in specific regions. For example, intraplate seismicity may be associated with deep features in the lithosphere and not known geological features at the surface (Assumpção et al. 2004). This second explanation for intraplate earthquakes includes density contrasts and heat flow (Mazabroud et al. 2005), buried rift pillows (Zoback and Richardson 1996), contrast in heat flow between continental and oceanic crusts (Sandiford and Egholm 2008), and glacial unloading (Jacobi et al. 2007).

The causes of the intraplate earthquakes in Northeastern Brazil are not completely understood, but one fundamental question remains if these events reactivate preexisting faults or are new structures. In Northeastern Brazil, Sykes (1978) was the first to propose that reactivation of major faults in the continent and transform faults in the ocean generate intraplate seismicity. Several other studies investigated this relationship. As described above, Precambrian ductile shear zones such as the Pernambuco lineament are structures that concentrate seismicity. Other structures such as the Patos shear zone present the same orientation in relation to the present stress field but exhibit no seismicity. A question remains why some shear zones have been reactivated by the present-day seismicity and others are not. Therefore, the complexity of this seismicity and its relationship with known structures deserve further investigations.

We present below a few cases of seismicity that both reactivate and cut across the preexisting fabric and faults in the basement. The examples are from the Pernambuco Lineament and at the border of the Potiguar Basin that indicate the complexity of the occurrence of seismogenic faults.

The Pernambuco Lineament, a ductile shear zone 700 km long and formed during the Brasiliano orogeny (740–540 Ma), is the best example of a major tectonic structure reactivated by the present-day seismicity. Seismicity in the lineament area is known since the nineteenth century. A seismic sequence in 1967 caused panic in the population and some inhabitants fled the city of Caruaru. The seismic activity started again in 1984 and has continued until the present. This shear zone and other in the region were reactivated during the breakup between South America and Africa in the Cretaceous (Matos 1992; Lima et al. 2016). The major event that struck the lineament occurred in 2006 and had a regional magnitude of Mb = 4.0. Either of the earthquakes reactivates the main E-W-striking shear zone, where they form normal faults, or they reactivate NE-SW-striking branches, where they form strike-slip faults (Ferreira et al. 2008, Lopes et al. 2010; Neto et al. 2013, 2014). In the southern block of the lineament, however, a recent study indicates that seismicity cut across the preexisting ductile shear zones (Neto et al. 2014).

The best-investigated intraplate seismicity in Brazil occurred in the Potiguar Basin to the east of the city of João Câmara. The seismicity formed a cluster of more than 40,000 events with magnitudes up to 5.1 Mb aligned along a NE-striking fault named as the Samambaia fault (Ferreira et al. 1987; Takeya et al. 1989; Bezerra et al. 2007). The events damaged more than 4,000 houses for over 8 years. A geometric and kinematic analysis of the events and focal mechanisms indicate that the Samambaia fault is a dextral, en echelon left-bend structure composed of at least three main segments. The surface projection of the fault coincides with a swarm of quartz and chalcedony veins, which indicate that this seismicity reactivates an existing fault along which fluid flow was important (Bezerra et al. 2007).

However, several seismogenic faults in the Borborema Province has no relationship with surface structures. For example, at the west part of the Potiguar Basin, however, the Cascavel fault cut across the preexisting fabric. This fault corresponds to a 5.2 Mb event that took place in November 1980 followed by aftershocks for more than two years (Berrocal et al. 1984; Ferreira et al. 1998). Two fault segments that represent events in 1980 and 1994 are 10 km apart (Fig. 1.3), but it is not clear if these are part of the same fault (Ferreira et al. 1998). The seismicity alignment cut across the Precambrian Basement fabric, indicating that this is possibly a new structure (Bezerra et al. 2011).

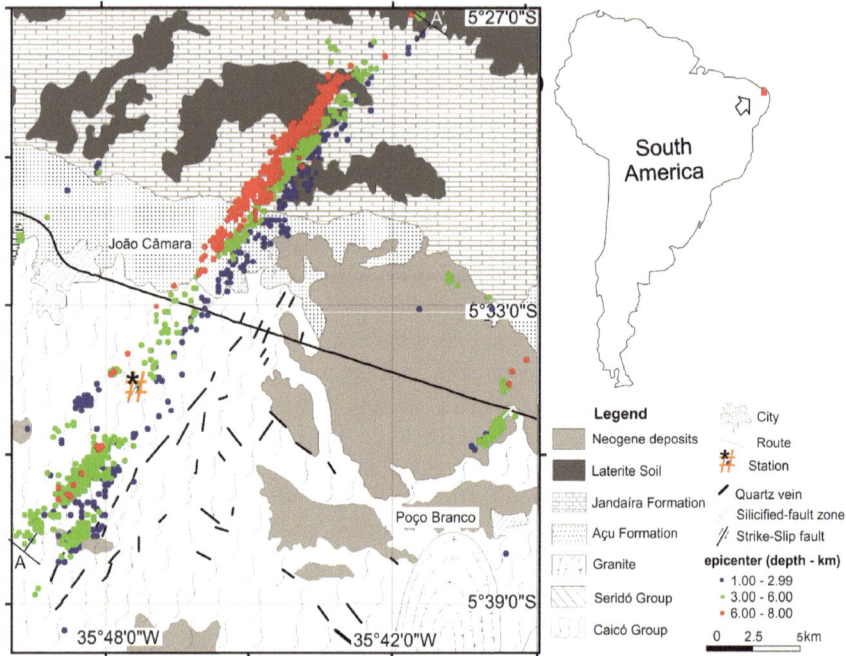

Fig. 1.3 Seismicity of the Samambaia fault in the eastern part of Northeastern Brazil. Epicenters are from the selected telemetric-network dataset (Laboratory of Seismology at the Federal University of Rio Grande do Norte) (modified from Bezerra et al. 2007)

1.5   Present-Day Stress Field

The stress field is the sum of all stresses for all points in a body of rock. The stress is the amount of force that acts in a given area of the rock and is expressed in Pascals. Three stress axes represent the stress field: $\sigma 1 \geq \sigma 2 \geq \sigma 3$. The present-day stress field in Northeastern Brazil was determined by the use of focal mechanisms, borehole breakouts, image logs, and geological indicators. These data indicate that the region is under a strike-slip stress regime with a general subhorizontal E-W-oriented compression and subhorizontal N-S-oriented extension (Assumpção 1992; Ferreira et al. 1998; Bezerra et al. 2011; Reis et al. 2013).

We present the results of each one of the stress indicators below.

1.5.1   Focal Mechanisms

Focal mechanisms represent the stress released by earthquakes using the stereographic projection. The directions of maximum and minimum compressive stresses and the possible direction of fault movement (slip direction) in focal mechanisms can be determined using the identification of the first movement of Primary P waves recorded in seismograms. The result of the method is a focal sphere surrounding the source of the earthquake, known as the beach ball, which represents the directions of maximum extension and compression during the event (Fig. 1.5). These directions may correspond to the main directions of tectonic stress, but may also deviate from these directions (e.g., Stein and Wysession 2003).

Focal mechanisms are the most widespread present-day stress field indicator in the region. Most of the stress field determination using focal mechanisms were carried in areas of local seismicity, during the monitoring of aftershocks (Ferreira et al. 1998, 2008). In general, focal mechanisms in the region indicate a strike-slip faulting regime, where the maximum ($\sigma 1$) and minimum ($\sigma 3$) stress axes are horizontal and the intermediate stress ($\sigma 2$) is vertical (Fig. 1.1). The existence of reverse and normal faulting focal mechanisms in a few sites is also consistent with an overall strike-slip stress regime. Overall, the maximum horizontal compression is roughly subhorizontal and trends E-W and the maximum horizontal compression is subhorizontal and trends N-S (Assumpção 1992).

The most accepted hypothesis is that the stress field in the region is a combination of local factors and regional factors. The most important regional causes are the ridge push in the mid-Atlantic ridge and slab pull in the Andes. One of the local causes is the density contrast between the oceanic and continental crust, which favors flexural bending and extension orthogonal to the continental margin (Assumpção 1992; Ferreira et al. 1998, 2008). For example, the maximum horizontal compression (SHmax) trends roughly E-W in the eastern part of the Potiguar Basin (Takeya et al. 1989; Bezerra et al. 2007), and bends to NW-SE in the central and western parts of the basin. Therefore, the maximum horizontal compression derived from the focal mechanisms is strongly influenced by and follow the continental margin (Ferreira et al. 1998). In the west part of the Borborema Province, the stress field is more scattered, but the overall SHmax still trends NW-SE. However, SHmax trends ESE-SSW along the Pernambuco Lineament. This variation in the trend of the SHmax is not observed with depth, i.e., in the same sites, the shallow and deep SHmax coincide from 1 to 12 km depth (Ferreira et al. 1998).

1.5.2 Borehole Breakouts

Breakouts are persistent areas of failure of borehole walls, which could persist for several hundreds of meters and result in oval shape distortion of the well in plan view. The largest diameter of the wall caused by the failure of the walls coincides with the direction of the minimum horizontal stress (Shmin), which is orthogonal to the maximum horizontal stress direction (SHmax) (Bell 1990).

In Northeastern Brazil, breakouts are concentrated along the continental margins in the Recôncavo, Sergipe-Alagoas, Potiguar and Ceará Basins. SHmax trends roughly parallel to the margin in the Recôncavo and Potiguar Basins, and is scattered

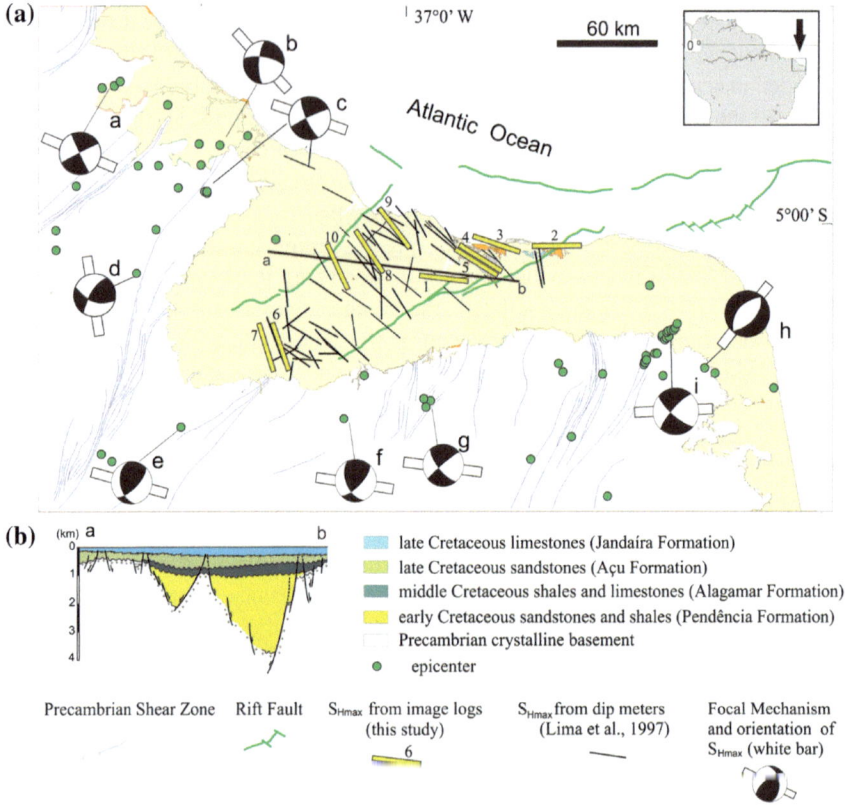

Fig. 1.4 Schematic principle of the right-diedra method (modified from Angelier 1994)

in the Sergipe-Alagoas and Ceará Basin. In the Sergipe-Alagoas Basin, for example, SHmax is WNW-ESE-oriented in the northern part of the basin and is mostly NE-SW-oriented in the southern part of the basin. In general, the behavior of SHmax indicates that the continental margin influences the stress field in the continental margin (Lima et al. 1997). The directions of SHmax derived from breakouts are consistent with the SHmax derived from the focal mechanism in the Potiguar Basin (Lima et al. 1997; Ferreira et al. 1998) (Fig. 1.4).

1.5.3 Image Logs

Borehole Imaging Logs are methods used to produce centimeter-scale images of a borehole wall in 2D. These methods can quantify the stresses in sedimentary basins and are widely used in the oil industry (Zoback 2010). Image logs were used in the Potiguar Basin to identify variations of the stress field with depth. The stress field shifts from a normal stress field from the surface to ca. 2500 m to a strike-slip stress field from 2500 to 4000 m. This pattern of shallow normal faulting and deep strike-slip faulting is consistent with a roughly NW-SE-oriented compression and regional folding and tectonic inversion of the basin (Reis et al. 2013).

1.5.4 Fault-Slip Data in Quaternary and Miocene Sedimentary Units

Joints and faults are used to indicate the directions of the stress axis. Joints represent one of the best markers as they propagate orthogonal to the direction of minimum stress ($\sigma3$). Faults are also ideal markers when they exhibit striae along the fault plane and where the direction of movement can be determined (Angelier 1994).

Fault-slip data were determined in the Potiguar, Paraíba, Rio do Peixe, and Araripe Basins. Most of the stress data both in Miocene and Quaternary sedimentary units indicate a strike-slip stress field with roughly E-W compression and N-S extension. These data also agree with focal mechanisms away from the margin.

References

Almeida FFM, Hasui Y, Neves BBB, Fuck RA (1981) Brazilian structural provinces: an introduction. Earth Sci Rev 17:1–29

Angelier J (1994) Fault slip analysis and paleostress reconstruction. In: Hancock PL (ed) Continental deformation. Pergamon Press, Oxford, pp 53–100

Assumpção M (1992) The regional intraplate stress field in South America. J Geophys Res 97(B8):889–903

Assumpção M, Schimmel M, Escalante C, Barbosa JR, Rocha M, Barros L (2004) Intraplate seismicity in SE Brazil: stress concentration in lithospheric thin spots. Geophys J Int 159:390–399

Assumpçao M, Schimmel M, Escalante C, Roberto Barbosa J, Rocha M, Barros LV (2014) Intraplate seismicity in Brazil. Intraplate Earthquakes, vol 1, 1st edn. Cambridge: Cambridge University Press, pp 50–71

Barros LV, Assumpção M, Quintero R, Caixeta D (2009) The intraplate Porto dos Gaúchos seismic zone in the Amazon craton—Brazil. Ectonophysics 469:37–47

Bell JS (1990) Investigating stress regimes in sedimentary basins using information from oil industry wireline logs and drilling records. In: Hurst A, Lovell MA (eds) Geological applications of wireline logs, vol 48. Geological Society of London, Special Publication, London, UK, pp 305–325

Berrocal J, Assumpção M, Antezana R, DiasNeto CM, Ortega R, França H, Veloso JAV (1984) Seismicity of Brazil. IAG-USP—Brazilian Commission of Nuclear Energy (in Portuguese)

Bezerra FHR, Vita-Finzi C (2000) How active is a passive margin? Paleoseismicity in Northeastern Brasil. Geology 28:591–594

Bezerra FHR, Ferreira JM, Sousa MOL (2006) Review of Seismicity and Neogene tectonics in northeastern Brazil. Revista de la Asociación Geológica Argentina 16:525–535

Bezerra FHR, Takeya MK, Sousa MO, Nascimento AF (2007) Coseismic reactivation of the Samambaia fault. Tectonophysics 430:27–39

Bezerra FHR, Brito Neves BB, Corrêa ACB, Barreto AMF, Suguio K (2008) Late Pleistocene tec-tonic–geomorphological development within a passive margin—the Cariatá trough, northeastern Brazil. Geomorphology 97:555–582

Bezerra FH, do Nascimento AF, Ferreira JM, Nogueira FC, Fuck RA, Neves BB, Sousa MOL (2011) Review of active faults in the Borborema province. Intraplate South America Integration of seismological and paleoseismological data. Tectonophysics (Amsterdam) 510:269–290

Branner JC (1912) Earthquakes in Brazil. Bull Seismol Soc Am 2:105–117

Branner JC (1920) Recent Earthquakes in Brazil. Bull Seismol Soc Am 10:90–104

Brito Neves et al (2014) The Brasiliano collage in South America: a review. Brazilian J Geol 44(3). Print version ISSN 2317-4889. São Paulo July/Sept

Campbell DL (1978) Investigation of stress concentration mechanism for intraplate earthquakes. Geophys Res Lett 5:477–479

Crone JA, Machette MN, Bowman JR (1997) Episodic nature of earthquake activity in stable conti-nental regions revealed by paleoseismicity studies of Australian and North American Quaternary faults. Aust J Earth Sci 44:203–214

Crone AJ, De Martini PM, Machette MN, Okura K, Prescott JR (2003) Paleoseismicity of two historically Quiescent Faults in Australia: Implications for Fault Behavior in Stable Continental Regions. Bull Seismol Soc Am 93:1913–1934

de Almeida FF, de Brito Neves BB, Carneiro CD (2000) The origin and evolution of the South American platform. Earth Sci Rev 50:77–111

DeMets C, Gordon RG, Argus DF, Stein S (1990) Current plate motions. Geophys J Int 101:425–478

Dentith MC, Featherstone WE (2003) Controls on intra-plate seismicity in southeastern Australia. Tectonophysics 376:167–184

Dewey J, Lamb S (1992) Active tectonics of the Andes. Tectonophysics 205(1–3):79–95

do Nascimento AF, Bezerra FHR, Takeya MK (2004) Ductile Precambrian fabric control of seismic anisotropy in the Açu dam area, Northeastern Brazil. J Geophys Res 109:B10311. https://doi. org/10.1029/2004jb003120

Ferreira JM, Assumpção M (1983) Seismicity of Northeastern Brazil. Brazilian J Geophys 1:67–88

Ferreira JM, Oliveira P.T, Takeya MK Assumpção M (1998) Superposition of local and regional stresses in northeast Brazil: evidence from focal mechanisms around the Potiguar marginal basin. Geophys J Int 134:341–355

Ferreira JM, Franca GS, Vilar CS, Nascimento AF, Bezerra FHR, Assumpcao M (2008) The role of Precambrian mylonitic belts and present-day stress field in the coseismic reactivation of the Pernambuco lineament, Brazil. Tectonophysics 456:111–126

Gordon RG, Jurdy DM (1986) Cenozoic global plate motions. J Geophys Res Solid Earth. https:// doi.org/10.1029/JB091iB12p12389

Gurgel SP, Bezerra FH, Corrêa AC, Marques FO, Maia RP (2013) Cenozoic uplift and erosion of structural landforms in NE Brazil. Geomorphology (Amsterdam), 186:68–84

Hanks TC, Johnston A (1992) Common features of the excitation and propagation of strong ground motion for Noth American earthquakes. Bull Seismol Soc Am 82:1–23

Jacobi RD, Lewis CMF, Armstrong DK, Blasco SM (2007) Popup field in Lake Ontario south of Toronto, Canada: Indicators of late glacial and post-glacial strain. In: Stein S, Mazzotti S (Eds.), Continental intraplate earthquakes: science, hazard, and policy issues: Geological Society of America Special Paper, vol 425, pp 129–147

Japsen P, Bonow JM, Green PF, Cobbold PR, Chiossi D, Lilletveit R, Magnavita LP, Pedreira A (2012) Episodic burial and exhumation in NE Brazil after opening of the South Atlantic. Geol Soc Am Bulletin. Published online on 13 January 2012 as https://doi.org/10.1130/b30515.1

Johnston AC (1996) Seismic moment assessment of earthquakes in stable continental regions. Geophys J Int 124:381–414

Lima CC, Nascimento E, Assumpção M (1997) Stress orientations in Brazilian sedimentary basins from breakout analysis: implications for force models in the South American plate. Geophys J Int 130:112–124

Lopes AE, Assumpção M, Do Nascimento AF, Ferreira JM, Menezes EA, Barbosa JR (2010) Intraplate earthquake swarm in Belo Jardim, NE Brazil: reactivation of a major Neoproterozoic shear zone (Pernambuco Lineament). Geophys J Int 180:1303–1312

Mascle J (1976) Atlantic-type continental margins: Distinction of two basic structural types. Anais da Academia Brasileira de Ciência. 48:148–155

Matos RMD (1992) The Northeast Brazilian Rift system. Tectonics 11(4):766–791

Minster JB, Jordan T (1978) Present day plate motions. J Geophys Res Atmos 83(B11). https://doi.org/10.1029/JB083iB11p05331

Neto HCL, Ferreira JM, Bezerra FH, Assumpção MS, do Nascimento AF, Sousa MO, Menezes EA (2013) Upper crustal earthquake swarms in São Caetano: Reactivation of the Pernambuco shear zone and trending branches in intraplate Brazil. Tectonophysics (Amsterdam), 608:804–811

Neto HCL, Ferreira JM, Bezerra FH, Assumpção M, do Nascimento AF, Sousa MO, Menezes EA (2014) Earthquake sequences in the southern block of the Pernambuco Lineament, NE Brazil: Stress field and seismotectonic implications. Tectonophysics (Amsterdam), 633:211–220

Reis ÁF, Bezerra FH, Ferreira JM, Nascimento AF, Lima CC (2013) Stress magnitude and orientation in the Potiguar Basin. Brazil: implications on faulting style and reactivation. J Geophys Res Solid Earth 1:n/a-n/a

Sandiford M, Egholm DL (2008) Marginal controls on Australian intraplate seismicity: questioning the role of structural reactivation. Tectonophysics 457:197–208

Schulte SM, Mooney WD (2005) An updated global earthquake catalogue for stable continental regions: reassessing the correlation with ancient rifts. Geophys J Int 161:707–721

Shedlock KM, Tanner JG (1999) Seismic hazard map of the western hemisphere. Ann Geophys 42(6)

Stein S, Wysession M (2003) An introduction to seismology, earthquakes, and earth structure. vol xi, p 498. Oxford: Blackwell Science. ISBN 0 865 42078 5

Stevenson D, Gangopadhyay A, Talwani P (2006) Booming plutons: source of microearthquakes in South Carolina. Geophys Res Lett 33:L03316. https://doi.org/10.1029/2005GL024679

Sykes LR (1978) Intraplate seismicity, reactivation of preexisting zones of weakness, alkaline magmatism, and other tectonism postdating continental fragmentation. Rev Geophys Space Phys 16:621–688

Talwani P (1988) The intersection model for intraplate earthquakes, January 1988. Seismol Res Lett 59(4):305–310

Zoback MD (1983) Intraplate earthquakes, crustal deformation and in-situ stress. In: Hays WW, Gori PL, Kitzmiller CA (Eds.), The 1886 Charleston, South Aarolina, earthquake and its implication for today. Open-File Report no. 83–843. US Geological Survey, Reston, VA, pp 169–178

Zoback ML, Richardson RM (1996) Stress perturbation associated with the Amazonas and other ancient continental rifts. J Geophys Res 101:5459–5475

Chapter 2
Understanding the Geological Setting of Northeastern Brazil

Abstract The geological setting of northeastern Brazil comprises the Borborema Province and Cretaceous to Cenozoic sedimentary basins. The Borborema Province corresponds to a region ~900 km long and ~600 km wide in northeastern Brazil. It is composed of Neoproterozoic fold belts with Archean and Proterozoic basement inliers, which were deformed during the Brasiliano orogeny at ~750–500 Ma. The tectonic stability of this Province was achieved in the Cambrian. The Province was mostly deformed by ductile shear zones, which bound major tectonic terrains. The coastal area of the region and part of its interior comprise mainly Cretaceous sedimentary basins, which mark areas of rifting during the breakup of Pangea and the separation of the African and South American plates. The major faults of the region were formed during this period. These rift basins mostly formed along the continental margin as far as 300 km from the present coastline, which correspond to areas of extended crust and crustal thinning. Fault reactivation in the Cretaceous occurred mostly where shear zone forme terrain boundaries. Paleoseismological data indicate that earthquakes struck the region in the last ~100 ka. Earthquake at least 5.5–6.0 Ms induced soft-sediment deformation characterized mainly by dikes, dome-like load structures, and mixed layers that affect gravel and sand strata from 400 to 10 ka in alluvial valleys. Paleoseismic data also exhibit evidence of surface faulting that either occur on both favorably oriented and unfavorably oriented faults in response to the stress field. Marine terrace deposits that correspond to two sea-level highstands and oxygen-isotope stages (7c, 220–206 ka and 5e, 117–110 ka) mark sea-level changes and tectonics along the coast.

Keywords Borborema province · Tectonic · Sedimentary basins

2.1   Introduction

This chapter summarizes the geological background of the Borborema Province, the sedimentary basins and Neogene–Quaternary sedimentary units in northeastern Brazil, which is necessary to understand the structural landforms described in the next chapters. Several studies have described neotectonic and paleoseismic deformation in the region. The term neotectonics was initially proposed by Obruchev (1948) to define a new branch of geoscience that investigates active geological processes. Subsequently, neotectonics was used as equivalent to Cenozoic or Quaternary deformation (e.g., Jennings et al. 1975). However, there is a tendency to abandon the precise ages of the period encompassed by neotectonics. According to Pavlides (1989), there is no definition of neotectonics that is globally valid and the onset of the neotectonic period should correspond to the period when the last stress field was established in a certain region. We use the term neotectonics to describe deformation that affected Northeastern Brazil since at least the Miocene, when the same stress field is observed in Miocene to Quaternary units. Paleoseismicity is the earthquakes recorded in the geological record, mostly in the Holocene.

Paleoseismology is the study of past earthquakes using the geological record and it is used for events that occurred in the past few thousand years (McCalpin and Nelson 1996). The study of paleoseismology includes the investigation of fault scarps, landslides, tsunamis, and soil liquefaction in late Quaternary sedimentary deposits (McCalpin 1999). The study of neotectonics and paleoseismology encompasses several investigations tools as described in (Fig. 9) Vita-Finzi (1986). The past decades have seen an improvement in techniques used for paleoseismic investigation and an increasing interest in the behavior of seismogenic faults.

Paleoseismology has great importance for several reasons. In intraplate areas fault recurrence can range from 10,000 to 100,000 or more (Crone et al. 1997), which is a period much longer than the age of most human settlements. In addition, the historical and instrumental seismic record does not represent the long-term behavior of seismogenic faults. Despite the fact that seismogenic faults have been related to known geological features such as surface faults, this association in most cases have little practical use because intraplate areas exhibit many of these structures, of which a few have seismicity (Stein 2009). Therefore, both historical and instrumental seismological records are too short to assess the behavior of seismogenic fault (Ambraseys 1992).

2.2   The Borborema Province

The Borborema Province is composed of Precambrian crystalline units of Archean and Proterozoic Age, which is overlain by Cretaceous and Neogene–Quaternary sedimentary units located in the northeastern corner of intraplate South America (Almeida et al. 2000; Caby et al. 1991; Santos et al. 2010; Brito Neves et al. 2014).

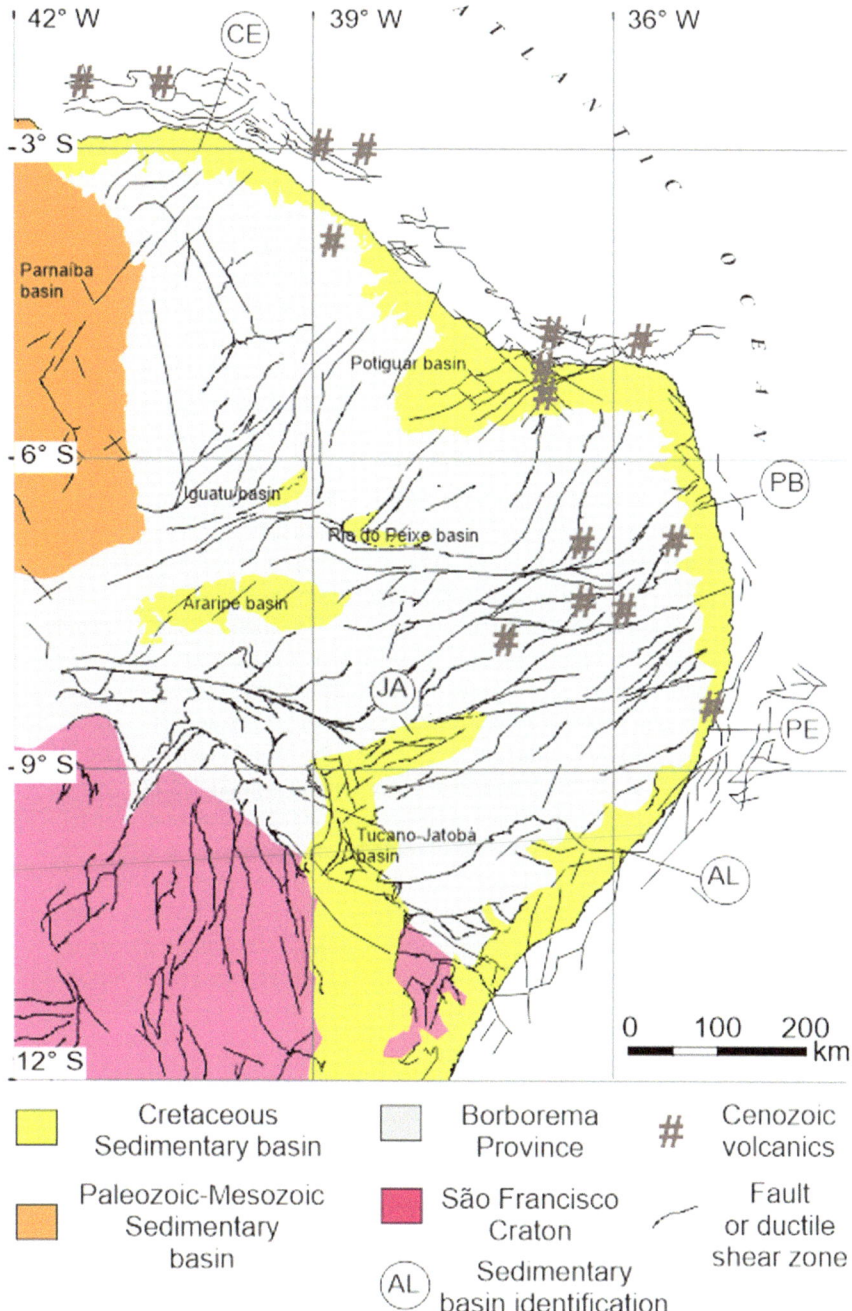

Fig. 2.1 Borborema province in Northeast Brazil

The Province is 900 km long along the N-S direction and 600 km wide along the E-W direction (Almeida et al. 2000). This province is limited to the west by the São Luiz-West Africa Craton and to the South by the São Francisco-Congo Craton (Fig. 2.1).

The Borborema Province covers an area of the northern Northeast located west of the Parnaíba Basin and north of the São Francisco Craton (Almeida et al. 2000) and is characterized by terrains of different lithologies separated by faults and lineaments (Brito Neves et al. 2000) with a predominant NE-SW and E-W orientation (Vauches et al. 1995). The Borborema Province is composed of various Achaean and Protero-zoic terrains assembling metamorphic and igneous lithologies. Several Paleozoic and Cretaceous sedimentary basins cap this Precambrian crystalline basement.

The Borborema Province represents an orogenic system affected by the deforma-tion, metamorphic, and magmatic events of Brasiliano (Pan-African) cycle between ca. 760 and 540 Ma (Almeida et al. 2000; Brito Neves et al. 2014). The peak of the deformation and metamorphism occurred at ~600 Ma and the tectonic stability of this province was achieved in the Cambrian. The Borborema Province is composed of Neoproterozoic terrains formed during the Brasiliano cycle and reworked Mid-dle to Paleoproterozoic and Archean terrains composed of volcanic, plutonic, and sedimentary units (Van Schmus et al. 2011; Brito Neves et al. 2014).

The most striking deformation feature of the Brasiliano cycle is a system of continental-scale steeply dipping shear zones, which is hundreds of kilometers long and 1–5 km wide. These shear zones mostly strike E-W and NE-SW, bound or cut across the major geological terrains of contrasting ages and lithologies. Some shear zones affect the whole lithosphere and reach the Moho (Tommasi and Vauchez 1997; De Castro et al. 2012). The shear zones continue in Africa in a pre-breakup reconstruction (Vauchez et al. 1995; De Castro et al. 2012) and are useful in matching up corresponding terrains in both continents separated during the breakup (e.g., Arthaud et al. 2008). A few studies identified links between both continents with some of these lineaments being more than 1000 km long along both continents (Van Schmus et al. 2011). The structural fabric in the geological terrains between the shear zones is complex and usually polyphasic, with several generations of metamorphic foliations (e.g., Santos et al. 2000; Brito Neves et al. 2014).

2.3 The Cretaceous Basins

After the Brasiliano cycle, the major deformation event was the diachronic rifting dur-ing the breakup between South America and Africa. Two types of rifting associated with two different kinds of margins developed during this breakup: (1) orthogonal extension and rifting along the N-S- to NE-SW-oriented east margin of South Amer-ica–Africa and (2) transform shearing along the E-W- to NW-SE-oriented equatorial margin of South America and Africa (Chang et al. 1992; Guiraud and Maurin 1992). The breakup progressed from south to north in the east margin and from west to east in the equatorial margin. Both rifting processes generated sedimentary basins on both plate margins and in the interior in the continental plates.

The NW-SE- to E-W-oriented extension of the crust favored the development of marginal and interior basins in Northeastern Brazil (Matos 1992, 2000). The basins include the marginal basins (Ceará, Potiguar, Paraíba, Pernambuco, and Sergipe-Alagoas) and the interior basins (Araripe, Iguatu, Rio do Peixe, Jatobá, Tucano). The area of these basins corresponds to an extended crust and crustal thinning (De Castro et al. 1998).

The rift phase of these basins was controlled by the reactivation of ductile Precambrian shear zones. These structures formed zones of weakness because they bound contrasting lithologies (Brito Neves et al. 1984; De Castro et al. 2012; Bezerra et al. 2014). The shear zones were reactivated from the late Jurassic to the Cretaceous during the breakup of South America and Africa and continued to be reactivated in a post-rift period (Nóbrega et al. 2005; Bezerra et al. 2014).

2.4 The Neogene–Quaternary Sedimentary Covers

The post-Cretaceous sedimentary record in the area is mostly marked by two main periods of sedimentation: the Barreiras Formation in the Oligocene–Miocene and the continental and marine deposits in the Quaternary. The most important and widespread deposition of sediments after the Cretaceous occurred in the late Oligocene to early–middle Miocene and formed the Barreiras Formation. This is a transitional to the marine clastic sequence, which outcrops from the west part of the Amazon River mouth to the state of Rio de Janeiro along more than 4000 km of the littoral zone of Brazil (Rossetti et al. 2013). The top of the Barreiras Formation is marked by a lateritic paleosoil formed from the the middle–late Miocene to the Pleistocene (Rossetti et al. 2012). There was a sediment deposition steadiness from the middle–late Miocene until the Pleistocene in the continental area (Rossetti et al. 2013).

Quaternary alluvial, Aeolian, and marine deposits occur in the region and are important markers of Neotectonic activity (Fig. 2.2). In the semiarid continental area, most of the alluvial sediments were transported by seasonal flash floods and formed braided deposits with mostly clast-supported conglomerates in a sand matrix (Rossetti et al. 2011). Mud- and silt-rich sediments occur in flood plains as a result of sedimentation or weathering of feldspar. These deposits range in age from 400 ka to the present along a few alluvial valleys such as the Assu Valley (Moura-Lima et al. 2011a, b).

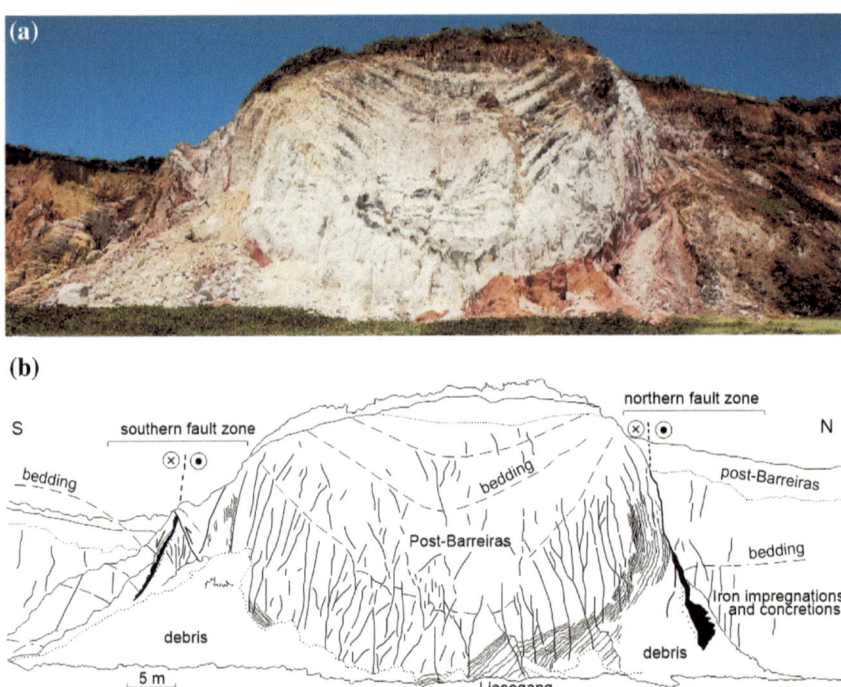

Fig. 2.2 The Miocene marine Barreiras Formation capped by Quaternary sediments. The latter forms a growth syncline, where fold is developed during sediment deposition (sea cliff south of João Pessoa City in the Paraíba Basin)

2.5 Introduction to Neotectonics in Northeastern Brazil

The influence of neotectonic faulting and folding in shaping landforms and controlling erosion and sediment deposition has been increasingly observed, but the level of tectonic stability of the region is still a matter of great debate. The low numbers of active fault escarps observed so far in Northeastern Brazil is probably due to the fact that fault slip rates are lower than erosion and sedimentation rates in the area (Nogueira et al. 2010; Bezerra et al. 2011). The long-term denudation rate has been estimated to be ca. 0.01 mm/year according to morphostratigraphic studies (Peulvast et al. 2006). However, in a few areas denudation rates range from 0.013 to 0.034 mm/year in uplifted plateaus such as Pereiro Massif (Gurgel et al. 2013).

2.6 Paleoseismicity in Northeastern Brazil

The historical and instrumental seismological records in Northeastern Brazil are too short to provide a complete evolution of the seismicity that could estimate a few parameters such as fault recurrence. Therefore, paleoseismicity provides a reliable approach for a more complete assessment of the seismicity.

We split the paleoseismic structures into two main categories: surface ruptures and liquefaction structures. One example of surface rupture occurs at the coastal plain close to the city of Natal. The Jundiaí fault controls the river valley and cut across both the Precambrian granites and sedimentary units from the Cretaceous to the Quaternary. The syn-sedimentary faults are part of the major Jundiaí fault. The SW end of the fault exhibits segments that trapped alluvial sediments during fault movement, which are ideal places to date fault activity. Luminescence optically stimulated ages using single-aliquot regenerative dose (SAR) chronology indicates that the fault was active in at least six periods in the past 100 ka, with a recurrence of ~16 ka. These ages could represent the maximum ages due to the poor bleaching of the sediments and it implies that the recurrence period could be underestimated. Radiocarbon ages of charcoal in faulted alluvial sediments yielded ages as recent as 4860–4570 (Bezerra and Vita-Finzi 2000). The surface rupture in the area is consistent with earthquake magnitude M ~ 5.3, which is closed to the magnitudes already observed in the instrumental record in the past 40 years (Nogueira et al. 2010).

Soft-sediment deformation structures are another kind of feature used in paleo-seismic studies in the region. These structures are mostly produced by liquefaction, which is a process whereby unconsolidated sediments from the solid state is caused to behave as a liquid due to the sudden increase in fluid pressure in the sediment matrix (Youd 1973). The threshold magnitude for liquefaction induced by earthquakes is 5.0 M (Obermeier 1996). This magnitude has been observed in Northeastern Brazil since the early nineteenth century (Ferreira and Assumpção 1983).

Soft-sediment liquefaction in gravels and gravelly sediments occurs in Quaternary alluvial deposits throughout the region. They are broadly distributed in several stratigraphic levels in alluvial channel deposits occurring along the fluvial valleys throughout the region. The most common type of structure in gravels is a combination of dikes, pillar, and pockets, where pebbles and cobbles of the host rock sank. Pockets form the base, pillar the middle, and pockets are the intermediate part of the system, which ranges in height from 1 to 4 m. These structure exhibit evidence of upward-directed water escape after deposition of sediment and before lithification (Saadi and Torquato 1992; Bezerra et al. 2005). Another type of structure also occurs in the region and forms sand volcanos, where the coarse nature of the sediments, their wide distribution across the basin in Quaternary sedimentary units, and their similarities with modern structures caused by earthquakes indicate a seismic origin (Fig. 2.3) (Moura-Lima et al. 2011a, b).

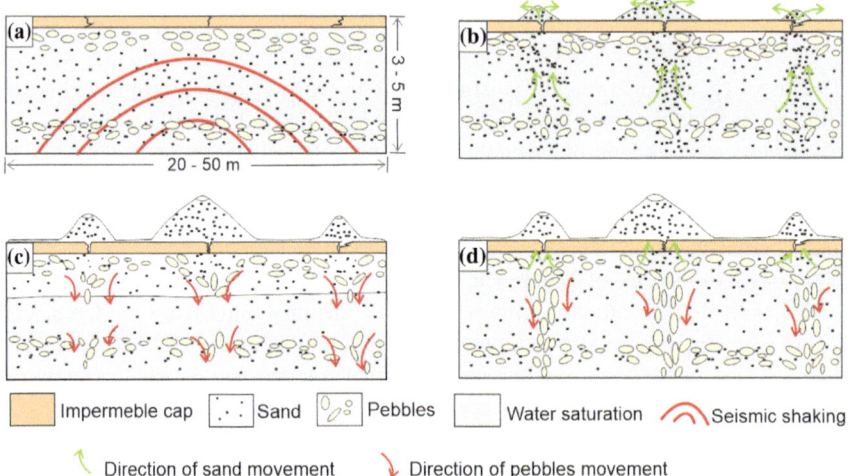

Fig. 2.3 Sediments liquefaction structure in gravels

Optically stimulated luminescence ages of the sediments indicate that these events occurred at least six times between ~352 and ~9 ka (Moura-Lima et al. 2011a, b). These structures are similar to dome-like load casts of convex upward lamination in a reverse density gradient system and the collapse of overlying gravels (Alfaro et al. 1997, 2010). Modern liquefaction in gravels is caused by earthquakes Ms ≥ 6.8–7.0 (Youd et al. 1985; Yegian et al. 1994), but they could also be caused by moderate earthquakes (Moura-Lima et al. 2011a, b). A non-seismic origin for some features cannot be ruled out.

Liquefaction in Quaternary sand layers has also occurred across the regions in layers that affect layers that affect almost entire basins such as the Paraíba. The soft-sediment deformation structures include isolated sand fragments in massive sandstones, sand dikes and diapirs, mixtures of plunged sediments, fitted sand masses, and sinks and bowls (Fig. 2.4).

These structures were formed contemporaneous or shortly after the deposition sediment deposition and before lithification and would have favored sediment accumulation and penecontemporaneous re-sedimentation. These features are confined to sharp-based stratigraphic horizons, which have the great regional extent of several hundreds of kilometer. These structures are similar to modern liquefaction caused by earthquakes at least Ms ≥ 5.0–6.0 (Rossetti et al. 2011b).

A combination of faulting and coeval soft-sediment deformation occurred in the São Francisco coastal plain between 82 and 8.3 ka, with small tectonic events after 1700 cal year BP. This coastal plain is bounded by normal faults striking N-S to NE-SW, which provided the accommodation space for deposition of Quaternary sediments in the coastal plain from the Cretaceous to the Quaternary. These faults have been described in seismic lines affecting Cretaceous strata (Destro et al. 1995;

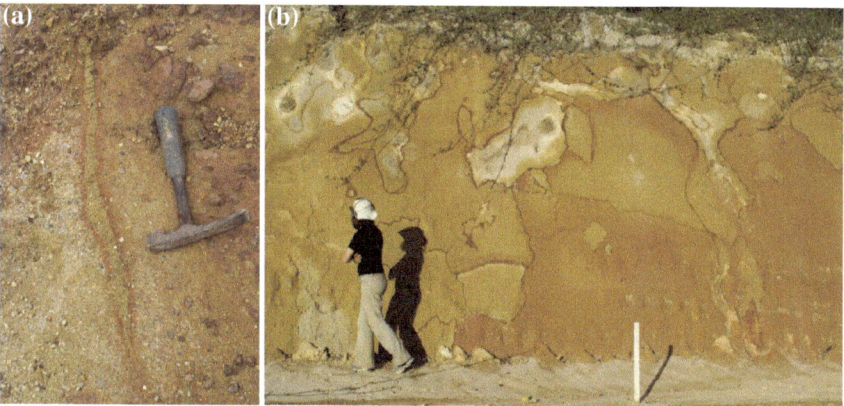

Fig. 2.4 Soft-sediment deformation in poorly lithified quaternary sediments in the Paraíba Basin: **a** sand dike and **b** mixtures of plunged sediments

Cruz et al. 2010). These faulting caused liquefaction of gravels and gravelly sediments and displaced Quaternary deposits, allowing the deposition of fan and coastal and marine deposits in the coastal plain (Lima et al. 2014).

2.7 Miocene to Quaternary Tectonics and Sea-Level Changes

The relative tectonic stability of the Atlantic coast of South America favors the reconstruction of sea-level episodes. The sea-level high stands caused widespread changes in coastal morphology, which greatly affected the ecological zonation of coastal systems and the creation of accommodation space for sediment deposition (Boski et al. 2015). Coastal and marine deposits have been increasingly investigated and have provided significant data as a basis for relative sea-level curves of different periods of the Earth history.

The use of coastal and marine deposits to identify crustal uplift and subsidence is also becoming increasingly common for several reasons. First, fossils and their traces occur frequently in marine deposits are, thus, easy to date. Second, sea-level high stands may form deposits that in a local scale may be considered horizontal and flat and could be used as a geometric and temporal marker. Third, crustal movements, even small ones, may perturb the delicate chemical, biological, and physical balance in coastal areas and could be easily identified (Stweart and Vita-Finzi 1998). In addition, progresses in glacial isostatic and field techniques have allowed the separation of the eustatic contribution of relative sea-level changes from other factors such as tectonics (Stewart and Vita-Finzi 1998). We present below studies that were carried out along the coastline in northeastern Brazil that shows the interplay between eustasy and tectonics in the Miocene, Pleistocene, and Holocene. We decided to describe the deposits of different ages separately due to the amount of different coastal uplift and subsidence that have affected these deposits.

The sea-level rises in the Oligocene and Miocene was the most significant since the end of Cretaceous, which flooded continental margins. Estimates indicate that these events amounted from 180 m (Haq et al. 1987) to a few dozens of meters (Miller et al. 2005).

The onshore part of the eastern continental margin of South America remained a non-depositional site from the end of the Cretaceous until the end of the Oligocene. During this time interval, the coastal was subjected to aerial exposure and the development of lateritic soil. Two transgressive sea-level episodes occurred after the period: one in the Oligo–Miocene and another in the early/middle Miocene. It caused the deposition of the Barreiras Formation. These characteristics make the Brazilian coast relevant to investigate global Oligocene and Miocene transgressions and their causes and related processes (Rossetti et al. 2013).

Several paleoenvironmental studies indicate that the Barreiras Formation was formed under the influence of tidal inner shelf and coastal marine processes (Rossetti et al. 1989; Rossetti and Góes 2009; Rossetti and Dominguez 2012). This unit was deposited in the Miocene mostly from 23 to 17 Ma according to Ar^{40}/Ar^{39} (Lima 2008) and (U-TH)/He ages (Lima 2008; Rossetti et al. 2013). These deposits are marked by a widespread unconformity in most sedimentary basins along the continental margin (Brown et al. 2000).

Field relationships of the Barreiras Formation indicate that the deposition of this unit was syn-deformation (Bezerra et al. 2001, 2014). Major normal and strike-slip faults commonly cut across this unit. Away from the major faults, the Barreiras Formation is gently folded into board synclines and anticlines (Fig. 2.5). At the outcrop scale, the unit is affected by vertical tectonic joints and by steeply dipping to low-angle normal and strike-slip faults.

Another evidence of synsedimentary deformation is the fact that the Barreiras Formation was deposited mainly in estuarine incised valley systems in the equatorial margin (Rossetti and Araújo 2004). It was also deposited in fault-controlled depocenters along the eastern continental margin, where the thickness of the unit increases from about 50 m to as much as 150 m and the base of the Barreiras Formation was offset by as much as 260 m (Bezerra et al. 2001; Nogueira et al. 2010). In addition, the age of deposition of this unit and its present-day elevation do not match any sea-level global curve, which strongly indicates that it has been subject to uplift after deposition. This uplift could either be caused by locally by tectonics and on by other factors such as dynamic topography on a more regional scale (Rossetti et al. 2013).

Two main sea-level high stands were dated along the Brazilian coast using geochronological methods. The first highstand is ~120 ka and was dated in the coastal of Central Brazil. The first ages correspond to eight samples of corals of the genus Siderastrea dated by Io/U. It yielded men ages about 123.5 Ma, which correspond with oxygen-isotope stage 5C, with elevations of 12 m $\pm$ 4 m asl (Bernat et al. 1983). This marine deposit was also identified by Barreto et al. (2002) and Suguio et al. (2011). The second high stand yielded optically stimulated luminescence ages

Fig. 2.5 Folds in the Miocene Barreiras Formation south of the city of João Pessoa—PB

~220 ka and corresponds to the 7c oxygen-isotope stage. Both high stands occur along the coastal area of the Pernambuco, Paraíba, and Potiguar Basins with the height varying from 0 to 20 m asl (Barreto et al. 2002; Suguio et al. 2011). The local offset observed in these deposits reach up to 20 m. In addition, these deposits cap folds and present liquefaction structures (Bezerra et al. 2011).

Ichnological and luminescence data along the Paraíba Basin also presented important tectonic results. Pleistocene deposits of nearshore paleoenvironment dated between 60.0 ($\pm$1.4) and 15.1 ($\pm$1.8) ka, when sea level was lower than the present. They exhibit trace fossils *Thalassinoides* and *Psilonichnus* at 38 m above present-day mean sea level. The occurrence of these marine units at this altitude cannot be explained by a rise in sea level. Although further tectonic, sedimentological, and geochronological data are needed, these first results indicate that tectonic uplift is the likely cause for such coastal uplift (Gandini et al. 2014).

Coastal deposits were dated along the coasts of the Potiguar, Paraíba, and Pernambuco Basins (Bezerra et al. 1998, 2003; Caldas et al. 2006; Suguio et al. 2013; Boski et al. 2015). These studies identified a rapid rise in sea level between 8300 and 7000 cal year BP at an averaged rate of approximately 6.1 mm/year^{-1}. The sea level approached the present-day position at 6700 cal year BP, followed by a subsequent high stand of 3–5 m asl at 5900–4500 cal year BP. Sea level then fell progressively, with possible oscillations in a few sites, after this period. This sea-level curve fits roughly the glacial isostatic model for the region (Peltier 1998).

Despite a few deposits are faulted, as a whole, the error caused by sea-level indicators in a mesotidal setting produced ambiguous results that do not allow sorting out the tectonic from the eustatic factors. This complexity is caused by the small offsets of Holocene deposits (usually less than 2–3 m) and the relatively high sea-level indicator errors in a mesotidal regime of more than 2 m of tide.

References

Alfaro P, Moretti M, Soria JM (1997) Soft-sediment deformation structures induced by earthquakes (seismites) in pliocene lacustrine deposits (Guadix-Baza Basin, Central Betic Cordillera). Eclogae Geologica Helvetica 90:531–540

Alfaro P, Gibert L, Moretti M, García-Tortosa FJ, Galdeano CS, Galindo-Zaldívar J, López-Garrido AC (2010) The significance of giant seismites in the Plio-Pleistocene Baza palaeo-lake (Spain). Terra Nova 22:172–179

Almeida FFM, Brito Neves BB, Carneiro CDR (2000) The origin and evolution of the South American Platform. Earth Sci Rev 50:77–111

Ambraseys N (1992) Soil mechanics and engineering seismology. In: Proceedings of 2nd Greek national conference of geotechnical engineer, Invited Lecture, Thessaloniki

Barreto AMF, Bezerra FHR, Suguio K, Tatumi SH, Yee M, Paiva R, Munita CS (2002) Late Pleistocene marine terrace deposits in northeastern Brazil: sea-level changes and tectonic implications. Palaeogeogr Palaeoclimatol Palaeoecol 179:57–69

Bezerra FH, da Fonseca VP, Vita-Finzi C, Lima-Filho FP, Saadi A (2005) Liquefaction-induced structures in quaternary alluvial gravels and gravels sediments, NE Brazil. In: Obermeier SF (ed) Paleoliquefaction and appraisal of seismic hazards. Eng Geol Amsterdam, vol 76, pp 191–208

Bezerra FHR, Vita-Finzi C (2000) How active is a passive margin? Paleoseismicity in Northeastern Brasil. Geology 28:591–594

Bezerra FH, Lima-Filho FP, Amaral RF, Caldas LH, Costa-Neto LX (1998) Holocene coastal tectonics in NE Brazil. In: Stewart I, Vita-Finzi C (eds.). Geological Society, London, Special Publications (Org). Coastal tectonics. 146 edn. Londres: Geological Society, vol 146, pp 279–293

Bezerra FH, Amaro VE, Vita-Finzi C, Saadi A (2001) Pliocene-quaternary fault control of sedimentation and coastal plain morphology in NE Brazil. J South Am Earth Sci 14:61–75

Bezerra FH, Barreto AMF, Suguio K (2003) Holocene sea-level history on the Rio Grande do Norte State coast. Brazil Marine Geol 196(1–2):15

Bezerra FH, do Nascimento AF, Ferreira JM, Nogueira FC, Fuck RA, Neves BB, Sousa MO (2011) Review of active faults in the Borborema Province, Intraplate South America Integration of seismological and paleoseismological data. Tectonophysics (Amsterdam) 510:269–290

Bezerra FH, Rossetti DF, Oliveira RG, Medeiros WE, Neves BB, Balsamo F, Nogueira FC, Dantas EL, Andrades Filho C, Góes AM (2014) Neotectonic reactivation of shear zones and implications for faulting style and geometry in the continental margin of NE Brazil. Tectonophysics (Amsterdam) 614:78–90

Boski T, Bezerra FH, de Fátima Pereira L, Souza AM, Maia RP, Lima-Filho FP (2015) Sea-level rise since 8.2 ka recorded in the sediments of the Potengi-Jundiai Estuary, NE Brasil. Marine Geol (Print) 365:1–13

Brito Neves BB, Fuck RA, Cordani UG, Thomaz Filho A (1984) Influence of basement structures on the evolution of the major sedimentary basins of Brazil: a case of tectonic heritage. J Geodyn 1:495–510

Brito Neves BB, Santos ED, Van Schmus WR (2000) Tectonic history of the Borborema Province, northeastern Brazil. In: Cordani UG, Milani EJ, Thomaz Filho A, Campos DA (eds) Tectonic evolution of South America. Rio de Janeiro, 31 International Geological Congress, pp 151–182

Brito Neves et al (2014) The Brasiliano collage in South America: a review. Brazilian J Geol 44(3). Print version ISSN 2317-4889. São Paulo July/Sept

Brown RW, Andrew JW, Gleadow (2000) Fission track thermochronology and the long-term denudational response to tectonics. In: Summerfield MA (ed) Geomorphology and Global Tectonics, Fission track thermochronology and the long-term denudational response to tectonics. Wiley

Caby R, Sial AN, Arthaud M, Vauchez A (1991) Crustal evolution and the Brasiliano Orogeny in Northeast Brazil. In: Dallmeyer R, Lecorché P (eds). Springer, pp 373–397

Chang HK, Kowsmann RO, Figueiredo AMF, Bender AA (1992) Tectonics and stratigraphy of the East Brazil rift system (EBRIS): an overview. Tectonophysics 213:97–138

Crone JA, Machette MN, Bowman JR (1997) Episodic nature of earthquake activity in stable continental regions revealed by paleoseismicity studies of Australian and North American Quaternary faults. Aust J Earth Sci 44:203–214

Cruz JB, Bento MD, Bezerra FHR, Freitas JI, Campos UP, Santos DJ (2010) Diagnóstico Espeleológico do Rio Grande do Norte. Revista Brasileira de Espeleologia 1:N1

Caldas LH, Stattegger K, Vital H (2006) Holocene sea-level history: evidence from coastal sediments of the northern Rio Grande do Norte coast, NE Brazil. Mar Geol 228:39–53

De Castro DL, Medeiros WE, Jardim de Sá EF, Moreira JAM (1998) Gravimetric map of the northeastern Brazil and adjacent continental margin: interpretation based on the isostatic hypothesis. Revista Brasileira de Geophysica 16:115–131

De Castro DL, Bezerra FH, Sousa MO, Fuck RA (2012) Influence of Neoproterozoic tectonic fabric on the origin of the Potiguar Basin, Northeastern Brazil and its links with West Africa based on gravity and magnetic data. J Geodynam 54:29–42

Destro N, Alkmim Fernando F, Magnavita Luciano P, Szatmari Peter (1995) The Jeremoabo transpressional transfer fault, Recôncavo–Tucano Rift, NE Brazil. J Struct Geol 25:1263–1279

Ferreira JM, Assumpção M (1983) Seismicity of Northeastern Brazil. Brazilian Journal of Geophysics 1:67–88

Gandini R, de Fátima Rossetti D, Netto RG, Bezerra FH, Góes AM (2014) Neotectonic evolution of the Brazilian northeastern continental margin based on sedimentary facies and ichnology. Quater Res (Print) 82:462–472

Guiraud R, Maurin JC (1992) Early Cretaceous rifts of Western and Central Africa: an overview. Tectonophysics 213(1):153–168. https://doi.org/10.1016/0040-1951(92)90256-6

Gurgel SP, Bezerra FH, Corrêa AC, Marques FO, Maia RP (2013) Cenozoic uplift and erosion of structural landforms in NE Brazil. Geomorphology (Amsterdam) 186:68

Haq BU, Hardenbol J, Vail PR (1987) Chronology of fluctuating sea levels since the Triassic. Science 235:1156–1167

Jennings CW (1975) Fault map of California. Calif Div Mines Geol Geol Data Map Ser Map. No. 1

Lima MD (2008) A história do intemperismo na Província Borborema Oriental, Nordeste do Brasil: implicações paleoclimáticas e tectônicas. 594 f. Tese (Doutorado em Geodinâmica; Geofísica) - Universidade Federal do Rio Grande do Norte, ... MariaGL_TESE.pdf, 18, 14

Lima CC, Bezerra FH, Nogueira FC, Maia RP, Sousa MO (2014) Quaternary fault control on the coastal sedimentation and morphology of the São Francisco coastal plain, Brazil. Tectonophysics (Amsterdam) 633:98–114

Matos RMD (1992) The Northeast Brazilian rift system. Tectonics 11(4):766–791

Matos RMD (2000) Tectonic evolution of the equatorial South Atlantic. In: Mohriak WU, Talwani M (eds). Atlantic rifts and continental margins. AGU Geophys Monograph 115:331–354

McCalpin JP, Nelson AR (1996) Introduction to Paleoseismology. In: McCalpin JP (ed) Paleoseismology. Academic Press, San Diego, pp 1–32

Moura-Lima EN, Sousa MO, Bezerra FH, de Castro DL, Damascena RV, Vieira MM, Legrand J (2011a) Reativação Cenozóica do Sistema de Falhas de Afonso Bezerra, Bacia Potiguar. Geociências (São Paulo) 30:77–93

Moura-Lima EN, Bezerra FH, Lima-Filho FP, de Castro DL, Sousa MO, Fonseca VP, Aquino MR (2011b) 3-D geometry and luminescence chronology of Quaternary soft-sediment deformation structures in gravels, northeastern Brazil. Sediment Geol 235:160–171

Neto HC, Ferreira JM, Bezerra FH, Assumpção M, do Nascimento AF, Sousa MO, Menezes EA (2014) Earthquake sequences in the southern block of the Pernambuco Lineament, NE Brazil: stress field and seismotectonic implications. Tectonophysics (Amsterdam) 633:211–220

Nóbrega MA, Sá JM, Bezerra FH, Neto JH, Iunes PJ, Guedes S, Saenz CT, Hackspacher PC, Lima-Filho FP (2005) The use of apatite fission track thermochronology to constrain fault movements and sedimentary basin evolution in northeastern Brazil. Radiat Measurem Amsterdam 39:627–633

Nogueira FC, Bezerra FHR, Fuck RA (2010) Quaternary fault kinematics and chronology in intraplate northeastern Brazil. J Geodyn 49:79–91

Obermeier SF (1996) Use of liquefaction-induced features for paleoseismic analysis—an overview of how their regional distribution and properties of source sediment

Obruchev VA (1948) Osnovnye cherty kinetiki i plastiki neotektonik. Akad Nauk Izv Serv Geol 5:13–24

Pavlides SB (1989) Looking for a definition of neotectonics. Terra Nova 1(3):233–235. https://doi.org/10.1111/j.1365-3121.1989.tb00362.x

Peltier WR (1998) Global glacial isostatic adjustment and coastal tectonics. In: Stewart I, Vita-Finzi C (eds), Coastal tectonics. Geological Society, London, Special Publications, vol 146, pp 1–29

Peulvast JP, Claudino-Sales V, Bezerra FHR, Betard F (2006) Landforms and neotectonics in the Equatorial passive margin of Brazil. Geodin Acta 19:51–71

Rossetti DF, Araújo J, Antonio ES (2004) Facies architecture in a tectonically-influenced estuarine incised valley fill of Miocene age, northern Brazil. J South Am Earth Sci 17, n. 4, pp 267–284

Rossetti DF, Dominguez JML (2012) Tabuleiros costeiros. In: Barbosa JSF, Mascarenhas JF, Gomes LCC, Dominguez JML, Souza JS (Org.). Geologia da Bahia: Pesquisa e Atualização. 1ed. Salvador: CBPM-UFBA 2:365–393

Rossetti DF, Góes AM (2009) Marine influence in the Barreiras formation, State of Alagoas, Northeastern Brazil. Anais da Academia Brasileira de Ciências (Impresso) 81:741–755

Rossetti DD, Góes AM, Truckenbrodt W (1989) Estudo paleoambiental e estratigráfico dos sedimentos barreiras e Pós-Barreiras na Região Bragantina, Nordeste do Pará.. Boletim do Museu Paraense Emílio Goeldi, BELÉM 1(1):25–37

Rossetti DF, Bezerra FH, Góes AM, Valeriano MM, Andrades-Filho CO, Mittani JC, Tatumi SH, Brito-Neves BB (2011a) Late quaternary sedimentation in the Paraíba Basin, Northeastern Brazil: landform, sea level and tectonics in Eastern South America passive margin. Palaeogeogr Palaeoclimatol Palaeoecol 191–204

Rossetti DF, Bezerra FH, Góes AM, Brito-Neves BB (2011b) Sediment deformation in Miocene and post-Miocene strata, Northeastern Brazil: evidence for paleoseismicity in a passive margin. Sed Geol 235:172–187

Rossetti DF, Góes AM, Bezerra FH, Valeriano MM, Brito-Neves BB, Ochoa FL (2012) Contribution to the stratigraphy of the onshore Paraíba Basin, Brazil. Anais da Academia Brasileira de Ciências (Impresso), 84:313–334

Rossetti DF, Bezerra FH, Dominguez JM (2013) Late Oligocene-Miocene transgressions along the equatorial and eastern margins of Brazil. Earth Sci Rev 123:87–112

Saadi, A., Torquato, J.R., 1992. Contribuição a neoetctônica do estado do Ceará. Revista de Geologia vol. 5. Universidade Federal do Ceará, pp 5–38

Santos EJ, Van Schmus WR, Kozuch M, Brito Neves BB (2010) The Cariris Velhos tectonic event in Northeast Brazil. J S Am Earth Sci 29(1):61–76. https://doi.org/10.1016/j.jsames.2009.07.003

Santos EJ, Ferreira CA, Silva Jr., JMF (2000) Geological map of the state of Paraíba. Brazilian Geological Survey—CPRM, Brazilian Ministry of Energy—MME, 1:500,00 Scale, Recife (In Portuguse)

Stein S, Liu M (2009) Long aftershock sequence within continents and implications for earthquake hazard assessment. Nature 462:87–89

Stewart IS, Vita-Finzi C (1998) Coastal tectonics. Geological Society Special Publication 146, The Geological Society, London, VIII + 378 p, 188 fig., index, 17.5 x 25.5 cm, 125 $. ISBN 1-86239-024-X

Suguio K, Bezerra FH, Barreto AM (2011) Luminescence dated Late Pleistocene wave-built terraces in northeastern Brazil. Anais da Academia Brasileira de Ciências (Impresso) 83:907–920

Suguio K, Barreto AM, Oliveira PD, Bezerra FH, Vilela MC (2013) Indicators of Holocene sea level changes along the coast of the states of Pernambuco and Paraíba, Brazil. Geologia USP. Série Científica 13:141–152

Tommasi A, Vauchez A (1997) Continental-scale rheological heterogeneities and complex intraplate tectono-metamorphic patterns—insights from a case-study and numerical modeling. Tectonophysics 279:327–350

Van Schmus WR, Kozuch M, Brito Neves BB (2011) Precambrian history of the Zona Transversal of the Borborema Province: insights from Sm-Nd and U-Pb geochronology. J S Am Earth Sci 31:227–252

Vauchez A, Neves S, Caby R, Corsini M, Egydio-Silva M, Arthaud M, Amaro V (1995) The Borborema shear zone system, NE Brazil. J S Am Earth Sci 8:247–266

Vita-Finzi C (1986) Recent earth movements: an introduction to neotectonics. Academic Press, p 226. ISBN-10:0127223703

Yegian MK, Ghahraman VG, Harutiunyan RN (1994) Liquefaction and embankment failure case histories, 1988 Armenia Earthquake. J Geotech Eng

Youd TL (1973) Liquefaction, flow, and associated ground failure. U.S. Geolog

Youd TL, Harp EL, Keefer DK, Wilson RC (1985) The Borah Peak, Idaho earthquake of Oct. 28, 1983. Liquefact Earthquake Spectra 2:71–98

Chapter 3
The Geomorphology of the Northeast: Classical and Current Perspectives

Abstract In the Northeastern Brazil, the relief documents important events of morphotectonic evolution and paleoclimate. Organized around paleosurfaces, the Northeastern Brazil presents several geomorphological compartments derived from major tectonic events, such as the Brasiliano Cycle and the separation between South America and Africa, in Cretaceous time. These events, printed in the relief on the contact with mountain-type morphology aligned according to different shear zones, lineaments and structural dissection, as well as drainage basins, were affected by Mesozoic uplift, which produced a complex system of morphostructures in the area. These features started to be interpreted in the 1960 as being formed by successive levels of paleosurfaces. Actually, the area displays a vast collection of structures and processes of Cenozoic age, including deformation, especially in sedimentary areas, which sometimes guide the work of external agents. In this context, this article discusses the models of geomorphological evolution of the area, particularly concerning paleosurfaces, analyzing the main characteristics and limitations in terms of its relation to Cenozoic tectonics and geochronology of geological units.

Keywords Northeastern · Morphology · Brazil · Planation surfaces · Morphostructural evolution of landscape

3.1 Classic Models of Geomorphological Evolution

The Northeastern Atlantic seaboard has a vast array of morphostructural compartments that were classically interpreted as the result of successive erosion cycles resulting from post-Cretaceous epeirogenic processes.

This geomorphological scenario was elaborated in morphostructural remnants of the Neoproterozoic orogeny named the *Brasiliano Cycle* (760–540 Ma, Brito Neves et al. 2014) that resulted in the Pannotia megacontinent. These kinds of landscapes developed along and close to shear zones that were reactivated during the Cretaceous and the Cenozoic, giving rise to arched or lowered areas, which once subjected to differential erosion, form an alignment of ridges and valleys with an NE-SW and E-W preferential direction. Currently, these uplifted and subsided areas are responsible for the formation of large morphostructural domains, in the Borborema Massif and its remnants, which together constitute the maximum arching of the northeastern shield (Saadi 1993).

When dealing with the geomorphological evolution of the Northeastern Brazil, it is possible to highlight that there are still few studies that deal with the origin and evolution of the landscapes through the lens of new the geomorphological knowledge. In this context, this chapter discusses the geomorphology of the region from the analysis of its classic models of evolution of erosion surfaces, their main limitations, and their relationship with the data referring to morphotectonics, evidenced by studies on the effects of Cenozoic tectonics on the evolution of the relief.

The study area covers Northernern Brazil and encompasses part of the northern area of Borborema Province. The relief in the area is arranged in the form of a vast eroded amphitheater facing the Atlantic and is marked by a central depression: the Jaguaribe depression, with a morphology inherited from morphostructural processes (Peulvast and Claudino Sales 2003) (Fig. 3.1).

Northeastern Brazil presents geomorphological compartments resulting from important tectonic events, such as the Brasiliano Cycle and the Cretaceous breakup

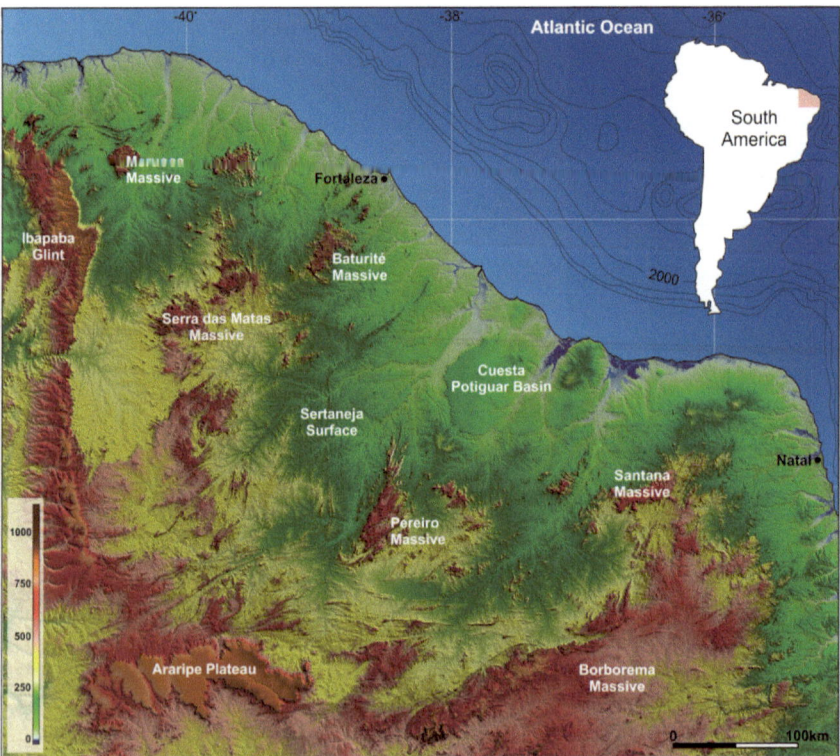

Fig. 3.1 The relief of Northeastern Brazil

between South America and Africa. These compartments make up the complex north-eastern geodynamic system and are impressed on the relief in the form of residual massifs aligned according to different shear zones, structural lineaments directing drainage and dissection, and Mesozoic Basins affected by uplift.

The relationship between erosive processes and uplifted blocks have been analyzed extensively in studies in the second half of the twentieth century (King 1956; Ab Sáber and Bigarella 1961; Bigarella 1994, 2003; Andrade and Lins 1965; Mabesoone and Castro 1975). In summary, these studies proposed a geomorphological organization in terraced levels of erosion surfaces. These levels would result from successive uplifts accompanied by phases of general erosion in dry climatic conditions or climatic alternations.

3.2 Genetic Aspects of the Relief of Northeastern Brazil: Classic Concepts

By analyzing the hydrographic network, various climatic variations and weathering profiles located at different coordinates, Dresch (1957) identified three leveled paleosurfaces at different coordinates. Demangeot (1960) identified four paleosurfaces attributing this to one erosive event succeeding each epeirogenic phase. Based on a study of geological/geomorphological profiles, Ab Sáber (1969) suggested the existence of five paleosurfaces in Northeastern Brazil resulting from a complex interaction between climatic changes and tectonic processes. These surfaces were formed by pedogenetic processes of humid climates alternated with morphogenetic phases of dry climates with sporadic and violent rainfall, where pediplanation processes prevailed. In this process, there is lateral retraction of the escarpments of the slopes and consequently detrital material accumulates at the base, forming gentle slopes toward the bottom of the valleys, called pediments. In the case of the climatic conditions remaining the same, there is a coalescence of the pediments and the formation of ample leveled surfaces called pediplanes (Fig. 4.13) (Fig 3.2).

The application of the aforementioned theory enabled the development of a Geomorphology of the Quaternary, with studies addressing the theme, without, however, there being a precise definition of the methodological treatment.

The model based on the occurrence of post-Cretaceous epeirogenies, accompanied by phases of dissection and pediplanation conducted by dry climates, was widely disseminated by Ab Sáber and Bigarella (1961), Bigarella (1994, 2003), Andrade and Lins (1965), Mabessone and Castro (1975), among others. These authors recognized the existence of various erosion surfaces, resulting from planation stages derived from erosion processes, following the uplift of a continental core. Thus, the sedimentary sequences of the Mesozoic and Cenozoic would be the result of erosion resulting from uplift and consequently the lowering of the regional base level.

The correlations between the continental and coastal deposits were analyzed by Fúlfaro and Suguio (1974). The interpretation and reconstruction of the sequence of

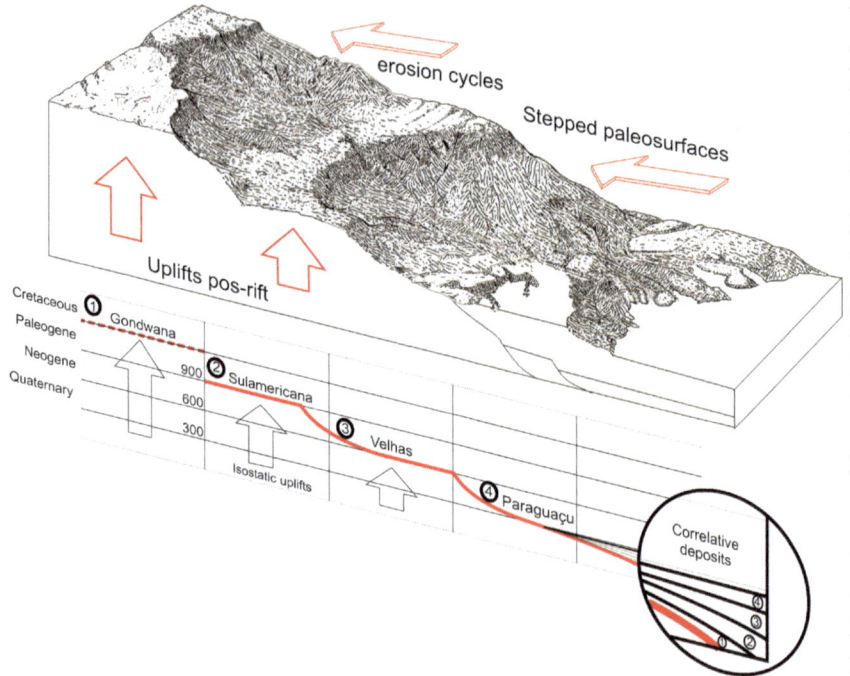

Fig. 3.2 Classical modelo about erosive surfaces in Brazil

Quaternary events were addressed by Tricart (1959) and Ab'Sáber (1969), whereas the evolution of the landscape through the chronology of the topographical forms was discussed by Bigarella and Andrade (1965). These studies corroborated the effects of the alternation of semiarid and humid phases. The formation of pediments would occur in the dry phases related to the glacial periods and low sea levels. These findings derived from the identification of deposits correlated to the phases of planation located below the current sea level. In these phases, the processes of pediplanation were associated to the retraction of the florets and the exposure of the soil formed during the previous humid phase. It follows that the regimen of sporadic rainfall promoted generalized erosion associated with lateral planation.

One example is the Sertaneja surface. Organized around the Borborema Plateau, the configuration of the relief makes this an important drainage disperser (Ab Sáber 1969), where a dense drainage network is responsible for intense dissection. Leveled areas between the elevated areas are formed where the denudation processes supplant aggradational ones, forming a vast erosional surface, also known as the *sertaneja depression* (Ab Sáber 1969).

As to tectonic activity, according to Saadi and Torquato (1994), the morphostructural evolution of the Northeastern Brazil was based on the occurrence of important crustal arching. Thus, the linear erosion processes sectioning the river valleys would be triggered by an uplift of polygenic origin. This process would create slopes that, when submitted to dryness, retreat laterally maintaining their altimetry, interpreted as paleosurfaces. The role of tectonics is evidenced in the sense of promoting the variations in the basement levels, inducing dissection.

This model is based on the interpretation of morphostructures as products of alternating periods of uplift (causing erosion), and stabilization (resulting in regional planation surfaces or erosion surfaces). In this context, the terraces, the planation surfaces, and the correlative deposits are essential data for the geomorphological analysis.

3.3 Synthesis of the Weak Points of the Paleosurface Model

From the discussion proposed herein, it appears that a major problem in evaluating the morphotectonic models suggested for Northeastern Brazil is the lack of geochrono-logical data to allow a better correlation between the proposed erosion cycles with the correlative sedimentation. This limitation stems from the fact that part of the sed-iments derived from the geomorphological evolution of the region is afossiliferous and therefore difficult to date chronologically.

Another noteworthy point is the difficulty in identifying and above all, correlating the paleosurfaces. The principle of a stepped relief with increasingly older surfaces located toward the crest does not apply widely in the region. In this case, aggradational paleosurfaces of similar geneses and ages are arranged at distinct elevations, as is the case of the Albian–Cenomanian surface identified by Peulvast and Claudino Sales (2003) in the post-rift section of the Potiguar Basin and the Araripe Basin (Fig. 4.6). The former not passing the height of 180 m and the latter reaching the height of 900 m.

Data on post-rift reactivation of faults and shear zones are not included in the pediplanation model and therefore neither is its effects on the geomorphology of the plains. However, various sites with faults affecting Cenozoic sedimentary units present a significant correlation with patterns of lineaments and drainage anomalies.

3.4 Current Concepts About the Geomorphology of Northeastern Brazil

With the advent and the consolidation of Morphotectonics, a Structural Geomor-phology began to emerge and gained significance in the work of Saadi et al. (1993, 1998, 2005), Peulvast and Claudino Sales (2000, 2003, 2006) and ultimately, in the world of current tectonics and its relationship with relief (Bezerra and Vita-Finzi 2000; Bezerra et al. 2008).

According to Bezerra et al. (2008), studies about the geomorphological evolution of the region for a long time have been based on a pediplanation model, where the morphology is a response to uniform uplift and the concomitant development of erosion surfaces. This conception is not confirmed if both the analytical and topographic criteria are morphostratographic and morphotectonic. There is growing evidence from morphotectonic indicators that the geomorphological evolution of Northeastern Brazil occurred in a much more complex manner than that proposed by the pediplanation model, which is limited in relation to the recent concepts regarding intraplate tectonics. This has happened because the pediplanation model does not incorporate rifting mechanisms and the history of basins; a limitation derived from the ideal of stability in intraplate areas such as the Brazilian territory. This model also fails to incorporate data on post-rift reactivation, and can be summarized as a model of uplift and planation that describes the passive equatorial margin of eastern South America and western Africa as successive erosion surfaces, developed due to an uplift and subsequent erosion.

Saddi (1993) and Peulvast and Claudino Sales (2003, 2005, 2006) question the model of successive post-Cretaceous uplifts, as being responsible for the development, until the Plio–Pleistocene, of successively embedded erosion surfaces. Some of these studies proposed a model in which the relief of the Borborema Province occurs around a central depression, the *Jaguaribe Depression* partially corresponding to the Cariri–Potiguar Jurassic–Cretaceous rift zone with a morphology characterized by segments of marginal escarpment, which is equivalent to the extremity of the aborted rift shoulders.

Peulvast and Claudino Sales (2003) reviewed the previously proposed erosion surfaces and recognized only three erosion surfaces. (1) *Sertaneja*, (2) *Jaguaribe* (Cenomanian), and (3) *Ibiapaba* (Paleozoic). They highlight the existence of ancient surfaces located in lowered elevations that contradicts the commonly accepted concept of steeped surfaces. These authors incorporated the tectonic processes regarding the Mesozoic breakup between South America and Africa and partly incorporated the subsequent Cenozoic tectonic activity. They also presented a proposal for the evolution of the relief of Northeastern Brazil, characterized by plateaus of a polygenic origin. In this context, they proposed that a continental flexure and thermal subsidence decisively contributed to the relief inversion in the Cenozoic.

Several studies proposed a chronology for the major morphostructural episodes responsible for the evolution of the relief:

1. (Early Cretaceous)

   1.1 Diffuse intracontinental extension (rifting), with the formation of rifts on the Cariri–Potiguar structural axis, with ac SE-NW orientation and a Neocomian age (145–130 Ma) (Ponte 1978, 1992; Bertanni et al. 1990; Matos 1992, 2000);

   1.2 Failed rifts in the Barremian (130–125 Ma), with the formation of the Araripe and Apodi sedimentary basins in the failed trenches (Bertanni et al. 1990; Matos 1992; 2000);

    1.3   The creation of transformational trenches through transtensional and transpressional efforts with an SE-NW orientation and E-W in the Aptian–Albian (between 125 and 100 Ma), creating fracture zones that have given rise to the Atlantic Ocean in the equatorial margin ca. 100 Ma (Matos 2000);

2. (Late Cretaceous)

    2.1   Thermal Subsidence of the sedimentary basins, allowing the deposition of the top layers between the Cenomanian and the Campanian (99–85 Ma), represented by the Açu and Jandaíra formations in the Potiguar Basin. These deposits considerably exceeded the limits of the basin, covering part of the adjacent leveled areas, which would therefore have been leveled in pre-Cenomanian periods (Matos 1992);

    2.2   Flexural uplift of the interior of the continent with the subsidence of the coastal area, ongoing to the present (Peulvast and Claudino Sales 2003).

3. (Paleogene–Neogene)

    3.1   Volcanism on the seaboard area (Macau, Ceará-Mirim) (Neogene, between 30 and 10 Ma) (Misuzaki et al. 2002);

    3.2   Eustatic variations, with the deposition of the Barreiras Formation in the Oligocene–Miocene (26–17 Ma) (Lima 2008; Rossetti et al. 2013). This would have been formed by continental sediments and probably transitional ones also, responsible for modeling the coastal plains; and the modeling of typical coastal formations (beaches, barriers, dunes, estuaries, lagoons, coastal plains) and river valleys;

    3.3   Inversion of sedimentary basins due to change in the stress field (Marques et al. 2013; Nogueira et al. 2015).

4. (Quaternary)

    4.1   Variations in the climate and sea level with the occurrence of a Holocene transgression and regression that modeled the coastal formations and the river valleys (Suguio et al. 1985, 2011, 2013; Bezerra et al. 1998, 2003; Barreto et al. 2002; Caldas et al. 2007).

    4.2   Seismcity and fault reactivation (Ferreira et al. 1998, 2008; Bezerra et al. 2011).

As a result of the events described above, a vast 450 km long amphitheater facing the sea was formed, which encompass parts of the Parnaíba Basin and the Borborema Massif. Nowadays, this amphitheater acts as a complex set of structures uplifted toward the south and greatly shaped by erosion (the Cariri/Potiguar rift zone) with tabular plateaus, sinking basins, and differential erosion corridors bordered by reliefs inherited from faults (Peulvast and Claudino Sales 2003).

In summary, residual crests aligned along the main tectonic directions, the borders of sedimentary plains affected by uplift, crystalline forms modeled by differential erosion, and residual massifs individualized by leveled depressions. In these areas, erosion processes supplanted aggradational ones and coastal plains shaped by eustasia, makes up the complex mosaic of the northeastern landscape and document important episodes of its morphotectonic and paleoclimatic evolution. Among these units, the coastal plains, river valleys, and coastal plateaux evidence important indicators of neotectonic events from various indicators. Given this scenario, recent studies have proven that intraplate seismic activity is an important mechanism for morphotectonic interpretation (Bezerra et al. 2007). For sedimentary areas, the effects of these paleostressors are expressed by different indicators.

Associated with neotectonic movement, different levels of gravel were identified in the valley of the Açu River (Fonseca and Saadi 1995; Moura-Lima et al. 2011) and in the Jaguaribe Valley (Maia 2005). Anomalies in the drainage of the Portalegre and Martins plateaus (Barros 1998) and the fault systems associated with them (Menezes 1999), such as the faults and deformations in the Barreiras Formation (Bezerra et al. 2001) and the liquefaction structures in the valley of the Açu River (RN) (Lima 2007; Moura Lima et al. 2011; Rossetti et al. 2011).

In the valleys of the Açu and Jaguaribe Rivers, deformations in neogenic sediments and evidence of the tectonic control of the morphology of features were identified by Maia (1993), (2005), Gomes Neto (2008), and Fonseca and Saadi (1995).

As an example of the neogene deformations associated with seismicity, Bezerra et al. (2005) identified numerous liquefaction structures in Quaternary sediments associated with intertwined channels of fluvial deposits in Rio Grande do Norte and Ceará. As to the occurrence of post-Pliocene tectonic activity, Bezerra et al. (2008) identified spasmodic colluviation processes associated with the reactivation of faults and the subsidence of grabens. According to luminescence data, such a reactivation would have occurred in two periods in the area of the Cariatá graben in Paraíba, namely: 224–128 and 45–28 ka.

However, it is worth noting that there are few works that deal with Cenozoic tectonics and their relationship to the relief, especially when it concerns their Cenozoic evolution.

From analyses carried out on the evolution of the relief in the Northeast of Brazil, the following factors are highlighted:

1. The model of the evolution of the northeastern landscape is based on the fact that the pedimentation ramps were submitted to climatic changes in the Quaternary and a terraced morphology, as a response to the epeirogenic tectonics. All the models are derived from evolution patterns of the relief with a tectonic (Davis 1899) or climatic (King 1960) focus. These classic models do not seem to match the reality observed in the group of geomorphological northeastern landscapes.

2. From various surveys, it is possible to point out with certainty that the erosion surfaces of continental dimensions are polygenic in origin (Peulvast and Claudino-Sales 2000).

3. Given the polygenic characteristics of the northeastern planation surfaces, it is suggested that the juxtaposition of several theories is a good alternative for a better understanding, considering that all dynamic factors involved in its modeling (back-wearing and down-wearing: Peulvast and Claudino-Sales 2000) are quite variable on the temporal–spatial scale.

References

Ab Sáber AN (1969) Participação das superfícies aplainadas nas paisagens do Nordeste Brasileiro. IGEOG-USP, Bol. Geomorfologia, SP, n 19, p 38

Ab Sáber AN, Bigarella JJ (1961) Considerações sobre a geomorfogênense da Serra do Mar. Boletim Paranaense de Geografia 4/5:94–110

Andrade GO, Lins R (1965) Introdução à morfoclimatologia do Nordeste do Brasil. Arquivos do Instituto de Ciências da Terra, Recife 3–4:11–28

Barreto AMF, Bezerra FHR, Suguio K, Tatumi SH, Yee M, Paiva R, Munita CS (2002) Late Pleistocene marine terrace deposits in northeastern Brazil: sea-level changes and tectonic implications. Palaeogeogr Palaeoclimatol Palaeoecol 179:57–69

Barros SDS (1998) Aspectos Morfo-Tectônicos nos Platôs de Portalegre, Martins e Santana/RN Dissertação de Mestrado PPGG–UFRN

Bezerra FH, Barreto AMF, Suguio K (2003) Holocene sea-level history on the Rio Grande do Norte State coast. Brazil Marine Geol 196(1–2):15

Bezerra FHR, Neves BBB, Correa ACB, Barreto AMF, Suguio K (2008) Late Pleistocene tectonic-geomorphological development within a passive margin—the Cariatá trough, northeastern Brazil. Geomorphology 01:555–582

Bezerra FHR, Takeya MK, Sousa MO, Nascimento AF (2007) Coseismic reactivation of the Samambaia fault. Tectonophysics 430:27–39

Bezerra FHR, Vita-Finzi C (2000) How active is a passive margin? Paleoseism Northeastern Brasil. Geology 28:591–594

Bezerra FHR, Fonseca VP, Vitafinzi C, Lima Filho FP, Saadi A (2005) Liquefaction-induced structures in quaternary alluvial gravels and gravels sediments, NE Brazil. Eng. Geol. In: Obermeier SF (ed), Paleoliquefaction and appraisal of seismic hazards, vol 76, pp 191–208

Bezerra FH, Lima-Filho FP, Amaral RF, Caldas LH, Costa-Neto LX (1998) Holocene coastal tectonics in NE Brazil. In: Stewart I, Vita-Finzi C (eds) Geological Society, London, Special Publications (Org.). Coastal tectonics. 146 ed. Londres: Geological Society, 1998, vol 146, pp 279–293

Bezerra FH, do Nascimento AF, Ferreira JM, Nogueira FC, Fuck RA, Neves BB, Sousa MOL (2011) Review of active faults in the Borborema Province, Intraplate South America Integration of seismological and paleoseismological data. Tectonophysics (Amsterdam) 510:269–290

Bezerra FHR, Amaro VE, Vitafinzi C, Saadi A (2001) Pliocene-Quaternary fault control of sedimentation and coastal plain morphology in NE Brazil. J South Am Earth Sci 14:61–75

Bigarella JJ (1994) Estrutura e Origem das Paisagens Tropicais, vol 1. Florianópolis: Ed. UFSC

Bigarella JJ (2003) Estrutura e Origem das Paisagens Tropicais, vol 3. Florianópolis UFSC

Brito Neves et al (2014) The Brasiliano collage in South America: a review. Braz J Geol 44:3. Print version ISSN 2317-4889. São Paulo July/Sept

Davis WM (1899) O Ciclo Geográfico. in: Geomorfologia—seleção de textos, vol 1 AGB USP, republicado em 1991. N 19

Demangeot J (1960) Essair sur le relief du Nord-est Brésilien. Ann. de Geographie, Paris 69(372):157–176

Dresch J (1957) Les problèmes géomorphologiques Du Nord-Est Brésilien. Bull Ass Géograp Français 263(264):48–59

Ferreira JM, Franca GS, Vilar CS, Nascimento AF, Bezerra FHR, Assumpcao M (2008) The role of Precambrian mylonitic belts and present-day stress field in the coseismic reactivation of the Pernambuco lineament, Brazil. Tectonophysics 456:111–126

Ferreira JM, Oliveira RT, Takeya MK, Assumpção M (1998) Superposition of local and regional stresses in northeast Brazil: evidence from focal mechanisms around the Potiguar marginal basin. Geophys J Int 134:341–355

Fonseca VP, Saadi A (1995) Compartimentos morfotectônicos no baixo curso do Rio açu (Açu-Macau), Rio Grande do Norte, in Simpósio de Geologia do Nordeste, 16, Recife: Boletim Sociedade Brasileira de Geologia, Recife-PE, vol 14, no. 1, pp 172–176

Fúlfaro VJ, Suguio K (1974) O Cenozóico paulista: gênese e idade. In: Congresso Brasileiro De Geologia, 28. *Anais*. Porto Alegre: SBG, vol 3, pp 91–101

King LC (1956) A Geomorfologia do Brasil Oriental. Revista Brasileira de Geografia, Ano XVIII(2)

Lima ME (2007) Liquefação em depósitos aluvionares do Rio Açu – RN. Dissertação de Mestrado, PPGG-UFRN, p 78p

Lima MD (2008) A história do intemperismo na Província Borborema Oriental, Nordeste do Brasil: implicações paleoclimáticas e tectônicas. 2008. 594 f. Tese (Doutorado em Geodinâmica; Geofísica)—Universidade Federal do Rio Grande do Norte, MariaGL_TESE.pdf, 18, 14

Mabesoone JM, Castro C (1975) Desenvolvimento geomorfológico do Nordeste Brasileiro. Boletim do Núcleo Nordeste da Sociedade Brasileira de Geologia. 3:3–5

Maia LP (1993) Controle Tectônico e evolução Geológica/Sedimentar da região da desembocadura do Rio Jaguaribe. Ceará, Dissertação de Mestrado, Departamento de Geologia, UFPE, Recife

Maia RP (2005) Planície Fluvial do Rio Jaguaribe: Evolução Geomorfológica. Ocupação e Análise Ambiental, Dissertação de Mestrado Geografia Física UFC Fortaleza - CE

Matos RMD (1992) The Northeast Brazilian rift system. Tectonics 11(4):766–791

Matos RMD (2000) Tectonic evolution of the equatorial South Atlantic. AGU Geophys Monogr 115:331–354. In: Mohriak WU, Talwani M (eds). Atlantic rifts and continental margins.

Menezes MRF (1999) Estudos sedimentológicos e contexto estrutural da Formação Serra dos Martins. Dissertação de Mestrado PPGG-UFRN

Peulvast JP, Claudino Sales V (2000) Dispositivos Morfo-Estruturais e Evolução Morfotectônica da Margem Passiva Transformante do Nordeste Brasileiro. III Simpósio Nacional de Geomorfologia, Campinas, SP

Peulvast JP, Claudino Sales V (2003) Stepped surfaces and Paleolandforms in the Northern Brasilian Nordeste: constraints on models of morfotectonic evolution. Geomorphology 3:89–122

Peulvast JP, Claudino Sales V (2005) Surfaces d'aplanissement et geodinamique. Géomorphologie (Paris) JCR, Paris, 4:249–274

Peulvast JP, Sales VC, Bezerra FH, Betard F (2006) Landforms and neotectonics in the Equatorial passive margin of Brazil. Geodinamica Acta (Paris), Paris (Submetido), 19:51–71

Rossetti DF, Bezerra FH, Góes AM, Brito-Neves BB (2011) Sediment deformation in Miocene and post-Miocene strata, Northeastern Brazil: evidence for paleoseismicity in a passive margin. Sed Geol 235:172–187

Rossetti DF, Bezerra FH, Dominguez JML (2013) Late Oligocene-Miocene transgressions along the equatorial and eastern margins of Brazil. Earth-Sci Rev 123:87–112

Saadi A (1993) Neotectônica da Plataforma Brasileira: Esboço de Intepretação preliminar. Geonomos MG. 1(1):1–15

Saadi A, Torquato JR (1994) Contribuição à neotectônica do Estado do Ceará. Revista de Geologia, Fortaleza-CE. 5:5–38

Saadi A, Bezerra FHR, Costa FD, Igreja HLS, Franzinelli E (2005) Neotectônica da plataforma Brasileira. In: Quaternário do Brasil. Holos Editora. São Paulo

Sousa MO, Bezerra FH, de Castro DL, Damascena RV, Vieira MM, Legrand JM (2011) Reativação Cenozóica do Sistema de Falhas de Afonso Bezerra, Bacia Potiguar. Geociências (São Paulo) 30:77–93

Suguio K, Bezerra FH, Barreto AM (2011) Luminescence dated Late Pleistocene wave-built terraces in northeastern Brazil. Anais da Academia Brasileira de Ciências (Impresso) 83:907–920

Suguio K, Barreto AM, Oliveira PD, Bezerra FH, Vilela MC (2013) Indicators of Holocene sea level changes along the coast of the states of Pernambuco and Paraíba, Brazil. Geologia USP. Série Científica 13:141–152

Tricart J (1959) Divisão morfoclimática do Brasil Atlântica Central. Boletim Paulista de Geografia, São Paulo 31:3–4

Chapter 4
The Erosion and Exhumation of Massifs in Precambrian Shear Zones

Abstract In northern Brazil, deformational structures control various features of the morphology. These structures are ductile shear zones and regional foliations formed during the brasiliano orogeny in the late proterozoic, and faults formed during the rifting of the margin in the Jurassic—cretaceous and in the late cretaceous and Cenozoic. The morphology features related to these structures include mainly residual alignment of ridges, fault-controlled valleys, and structural fault scarps. In this perspective, this paper presents a review about the morphostrutural control played by brittle and ductile basement structures on the relief in northeastern brazil. Our study indicates that continental denudation have been controlled by both differential erosion and tectonics. In this aspect, the exhumation along NE-SW and E-W striking shear zones originates lineaments, which confine the drainage channels and guide the dissection and sometimes fluvial aggradation. This process results in ridges and valleys that control the flow channels, which become indicators of shear zones and faults. This process is also observed in sedimentary basins, where dissection along fault zones form tectonic valleys that constitute the geomorphic expression of the fault reactivation.

Keywords Geomorphology · Northeastern Brazil · Morphotectonics

The geomorphology of Northeastern Brazil is notably marked by ductile and brittle deformation structures imprinted on Precambrian crystalline bedrock. These structures are represented by a set of inherited morphologies developed in fault zones of the Precambrian structure. Ductile structures such as the shear zones formed during the Brasiliano orogeny in the Neoproterozoic with a predominantly NE-SW and E-W orientation (Brito Neves 1999). These ductile structures such as shear zones were reactivated in a brittle form or new independent faults were formed in the Cretaceous during the separation of the Pangea megacontinent (Matos 1992; De Castro et al. 2012). The fault trends related to the ductile shear zones also have a NE-SW and E-W orientation (Bezerra et al. 2014; de Castro et al. 2008; De Castro et al. 2012).

Regarding the morphology, the shear zones exert an important control on the features that include structural massifs, linear crests, and incised valleys composing topographic highs and lows oriented according to the direction of the structural trends. The classical studies by Ab Sáber and Bigarella (1961), Bigarella (1994,

2003), Andrade and Lins (1965), and Mabessone and Castro (1975), interpreted this morphology as the result of successive erosion cycles induced by the lowering of the basement level, in turn, derived from post-Cretaceous epeirogenic processes. These studies were important contributions to the understanding of the geomorphological evolution with a morphoclimatic emphasis. However, these works do not consider the effects of the deformation of the deep crust and the exhumation of the ductile shear zones in the surface morphology. Thus, the role of tectonics in the evolution of the relief process is an essential factor in the evolutionary analysis, especially in areas of recurrent tectonic activity such as the Equatorial Atlantic seaboard of Brazil.

From the perspective of analyzing the role of morphotectonics in the genesis and evolution of the relief of northeastern Brazil, our study investigates the configuration of the model on a regional scale with an emphasis on a few structures, especially the exhumation of Precambrian strike-slip shear zones. For this purpose, we propose a macro compartmentation of the relief; this distinction is made for the geomorphological evolution of each morphostructural unit, as well as a description of the current relief and its association with Cenozoic morphodynamic aspects (Fig. 4.1).

The Brasiliano orogeny was responsible for the final amalgamation of eastern Gondwana Between 750 and 540 Ma (Arthaud 2007). At the end of this orogeny, the different crustal blocks appear to have been uplifted in a variable form along the main lineaments either as a result of reactivations of the extensive shear zones in a ductile–brittle fashion or associated to post-tectonic isostatic readjustments (Corsini et al. 1998). Subsequently, the phase of reactivation of the shear zones (early Jurassic to late Cretaceous), was characterized by the formation of rifts that resulted in the inland Reconcavo, Tucano, Jatoba, Araripe, Rio do Peixe, and Iguatu Basins and the coastal basins of the Atlantic seaboard such as Sergipe-Alagoas, Pernambuco, Paraíba, Potiguar, and Ceará. Many of the sedimentary deposits resulting from these successive tectonic events were influenced by underlying faults, particularly by the structures developed during the tectonics that were active in the late Brasiliano cycle until the Cretaceous (Matos 1992). These faults are pre-, syn- and post-rift (Bezerra and Vita-Finzi 2000; de Castro et al. 2000).

In this respect, the relief of the Borborema province corresponds to the set of highlands that are distributed along the eastern facade of Northeastern Brazil, north of the São Francisco River, above 200 m, whose boundaries are marked by a series of topographic leveling, and its epeirogenic genesis is associated with the fragmentation of Pangea and the active intraplate magmatism throughout the Cenozoic era (Correa et al. 2010).

From an uplift of polygenic origin, the linear erosion processes would be triggered, causing slopes that, subject to drought, would retreat laterally maintaining their altimetry, interpreted as paleosurfaces. The role of tectonics would be demonstrated in promoting the variations in the base levels, inducing dissection. The model is based on the interpretation of morphostructures as products of alternating periods of uplift (causing dissection) and stabilization (resulting in regional planation surfaces) (Maia et al. 2010). In this context, the terraces, the plantation surfaces, and the related deposits would be the source of essential data for geomorphological analysis.

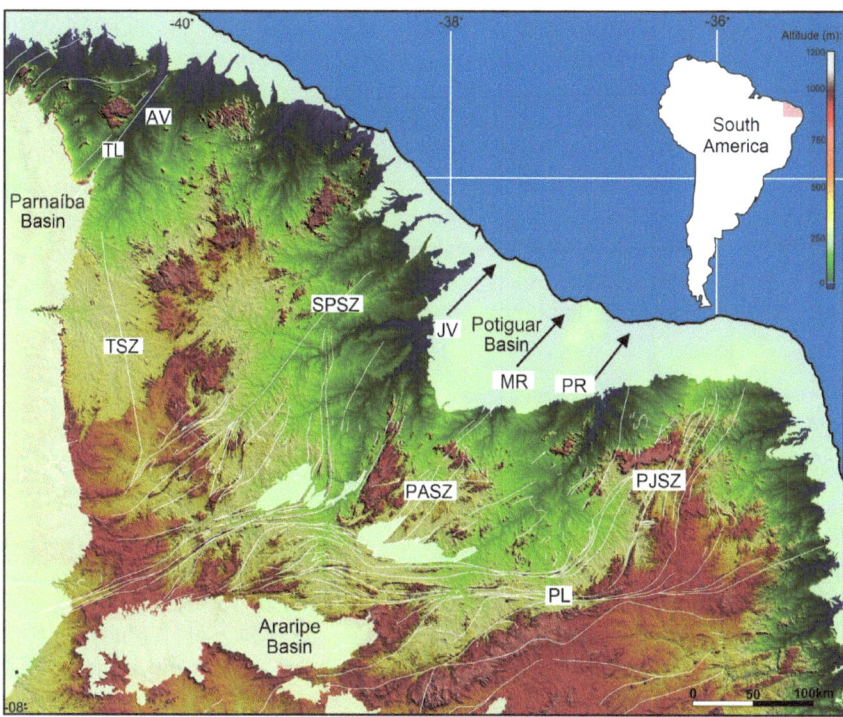

Fig. 4.1   The correlation between lineaments and the disposition of the relief of Northeastern Brazil. TL: Transbrasiliano Lineament, TSZ: Tauá Shear Zone, SPSZ: Senador Pompeu, JSZ: Jaguaribe, OSZ: Orós, PASZ: Portalegre, PJSZ: Picuí-João Câmara, PL: Patos Lineament JV: Jaguaribe River Valley, MV: Mossoró River Valley, PV: Piranhas River Valley (modified from Maia and Bezerra 2014)

The classic models and their morphoclimatic emphasis provided important support to the consolidation of a Quaternary Geomorphology. However, these works have limitations concerning the structural aspect, especially with regard to the effects that the tectonic rift, post-rift, and differential erosion developed along the shear zones have on the relief evolution.

The relief of Northeastern Brazil has been described classically as paleosurfaces resulting from peneplanation cycles (Maia et al. 2010). However, there is some evidence that the relief is controlled by ductile shear zones and their reactivations. Two aspects are important for this control. The first control is made by differential erosion. In this aspect, the shear zones are generally marked by granitic bodies, which are more resistant to erosion thereby forming a steep topography, which contrasts with the surrounding area. The second control results from the brittle reactivation of the ductile shear zones. This, in turn, has generated depressions and uplift according to the Brasiliano deformation planes thus causing a relief notably marked by structural lineament trends.

Sabins (1996) defined the term lineament as being a topographical linear feature that may represent a zone of structural weakness. These features are mappable and can be simple or compound, and their parts are aligned in a rectilinear manner or smoothly curved, reflecting subsurface phenomena (O'Leary et al. 1976). In geomorphological terms, lineaments represent common variations in the elevation of the land, the alignment of crests, segments of escarpments, stretches of drainage and valleys (Jordan and Schott 2005), which may indicate the location where geological structures that influence the relief evolution. Thus, the analysis of the domains of lineaments on a regional scale has been shown to be a useful tool to investigate the relationship between Geomorphology and Tectonics.

4.1 Influence of Fault Reactivations on the Relief Evolution in the Borborema Province

In evolutionary terms, the main events of tectonic evolution of the Borborema Province can be distinguished as follows: 1—Brasiliano Orogenesis derived from Brasiliano tectonic/ Pan-African bonding in 600 Ma (Brito Neves et al. 2000), which was coeval with an important granite plutonism (Almeida et al. 2000; Brito Neves et al. 2014). 2—Fragmentation of the Gondwana megacontinent resulting in the South America–Africa breakup (Matos 2000), and 3—Cenozoic tectonic reactivations (Bezerra and Vita Finzi 2000; Maia and Bezerra 2014), which influenced the geomorphological evolution through changes in baseline levels, inducing dissection and aggradation (Maia et al. 2010). The reader is referred to Sect. 4.3 of this book for details of events.

This sequence of tectonic events is the main cause of the master lines in the relief that controls the current geomorphological evolution, which is marked by the Cenozoic denudation processes. This control is the result of the structural conditioning of the geomorphological evolution, especially in areas of tectonic deformation such as shear zones. The shear zones are conduits through which a large volume of fluids can circulate, thus constituting areas that were the result of ductile deformation (Passchier et al. 1993). In these zones, the heterogeneous lithologic configuration is arranged in parallel strips orientated according to the foliations planes. The subsequent reactivation of these ductile belts generated systems of faults and joints that influence erosion processes. The anisotropy of shear zones and their role as weakness zones are also caused by the rise of granitic magma and its subsequent horizontal displacement occur, resulting in linear intrusions. This offset was evidenced by the parallelism that exists between the magmatic and mylonitic foliation (Neves 1991). In this way, the tectonic deformation zones created lineaments that correspond to mylonitic belts mainly formed along linear granitic intrusions.

The shear zones are formed at a deep crustal level; however, they were reactivated on various crustal levels. These reactivations formed faults on the ductile–brittle and brittle fields. Both processes are potential generators of lineament trends, which for geomorphology are considered important features related to tectonic deformation. Such deformations are associated with structures in the subsurface, and can be identified in remote sensing imagery and in the field.

Thus, it is important to highlight the structural nature of the relief, be it through the control exercised in the differential erosion due to the lithological heterogeneity associated with the ductile shear zones or through the effects of the brittle reactivations of these zones. In Northeastern Brazil, the Cenozoic tectonic reactivations are responsible for the rejuvenation of the crystalline massifs through their uplift or through the depression of blocks through subsidence that resulted in the formation of small interior basins. Regarding this, Gurgel et al. (2013) used morphotectonic analysis to propose a tectonic origin for the escarpments that sustain the Pereiro Massif, which is situated between the States of Ceará and Rio Grande do Norte, thus contrasting with the more traditional classifications that described it as a residual massif.

The effects of the tectonic reactivations on the morphology, drainage and sedimentation environments of different basins in the Equatorial margin were analyzed by Bezerra and Vita-Finzi (2000), Bezerra et al. (2001, 2008), Furrier et al. (2006), Nogueira et al. (2010), Moura Lima et al. (2010), Lima (2010), Rossetti et al. (2011), and Maia (2012). These studies pointed out that the relationship among Cenozoic tectonics, the Neogene Barreiras Formation, and Quaternary deposits is responsible for the conformity between the orientation of the valleys, cliffs, and neotectonic faults. According to these works, there is a genetic relationship between older alignments and the current morphology of the valleys and coastal cliffs. Several alignments of the valleys and depressed areas are positioned according to the orientation of the Precambrian and Cretaceous fault lines, which may represent recent reactivation of these lines of weakness (Moura Lima et al. 2010).

4.2 Structural Control of the Relief and Drainage: Examples from the Borborema Province

In Northeastern Brazil, the NE-SW- and E-W-trending lineaments often represent the surface expression of the Brasiliano deformation of a ductile/brittle nature that was reactivated in the Cretaceous and in the Cenozoic. These lineaments are trends of faults that have an important influence on the structural control on the Quaternary drainage, dissection of the topography and deposition of sediments. Examples of this morphostructural orientation are commonly found in the erosional areas represented by the *sertanejo* depression and the crystalline massifs.

The continental-scale transcurrent Brasiliano shear zones cut across the Borborema Province in E-W and NE orientations, where they generally control the intrusion of several granitic bodies (Nascimento 1998). It follows from this that the relief is commonly arranged in sequences of ridges and valleys oriented according to the lineament trends. These trends are represented by foliation planes, by quartzite crests, by granitic intrusions and mylonite foliations. This is the origin of the parallel lineaments with differentiated resistance to geochemical and physical denudation, favoring through differential erosion that wears of the weaker tracks according to the deformation planes and maintains intrusive bodies, which are then expressed in the relief as structurally controlled residual crests.

The erosion of the Precambrian ductile shear zones carries out a progressive exhumation of the crystalline massifs, traditionally described as residual. Through differential erosion, this exhumation enabled the formation of sets of geomorphological lineaments arranged in positive and negative linear forms following the trends of the Brasiliano foliation.

In this way, the Brasiliano foliations control diverse morphologies on a regional scale, especially those associated with the ductile shear zones, where the relief is characterized by symmetrical crests, where slopes present an accentuated declivity arranged in a continuous form. Examples of this contextualization include the NW sector of the Borborema Province in the state of Ceará close to the Transbrasiliano Lineament, the Cariri-Potiguar trend, the zone of the transversal domain and the northern sector of the Borborema Massif.

In the northern sector of the Borborema Massif, extensive elevations with a predominantly NE-SW orientation compose a group of massifs that are distinguished from the leveled topography of the *sertanejo* depressions, emerging as elevations with an average height between 400 and 700 m of altitude.

The structural massifs are important markers of the morphotectonic and morphoclimatic evolution, mainly arranged as linear crests or strongly desiccated incised valleys. Their geomorphological evolution is controlled by processes of differentiated erosion, due to their geological constitution being derived principally from intrusive plutonic rocks or metasedimentary rocks in the case of quartzite crests. As the latter is more resistant to erosion processes, they remain in the topography as lengthened crests in the direction of the structural trends or in the form of peaks, as in the case of volcanic intrusions.

Among these features, the Cabugi Peak is noteworthy for its conical morphology. It is an exhumed volcanic neck, currently about 590 m a.s.l.. From a geological point of view is the most recent record of Brazilian Cenozoic continental magmatism (Ferreira and Sial 2002) (Fig. 4.2).

The surrounding rocks are subjected to more strongly pronounced effects of dissection, which makes these massifs important testimonies of Cenozoic erosion.

Among the massifs in Rio Grande do Norte, the mountains of Portalegre, Martins (Block B, Fig. 3.2), João do Vale, and Santana (Block C, Fig. 3.2) have a peculiar topography, with flat tops, which gives them an atypical morphology in terms of the residual relief of the Brazilian Northeast. These flat surfaces that cap the massifs are composed of Cenozoic continental Paleogene-Neogene sediments (Figs. 4.3, 4.4 and 4.5).

These are plateaus around 700 m a.s.l., where the uplifted unconformity between Cenozoic sediments and the Precambrian crystalline basement is capped by the conglomeratic sediments of the Serra dos Martins Formation (Menezes 1999), of the

Fig. 4.2 Cabugi peak—volcanic neck (*Source* Maia et al. 2009)

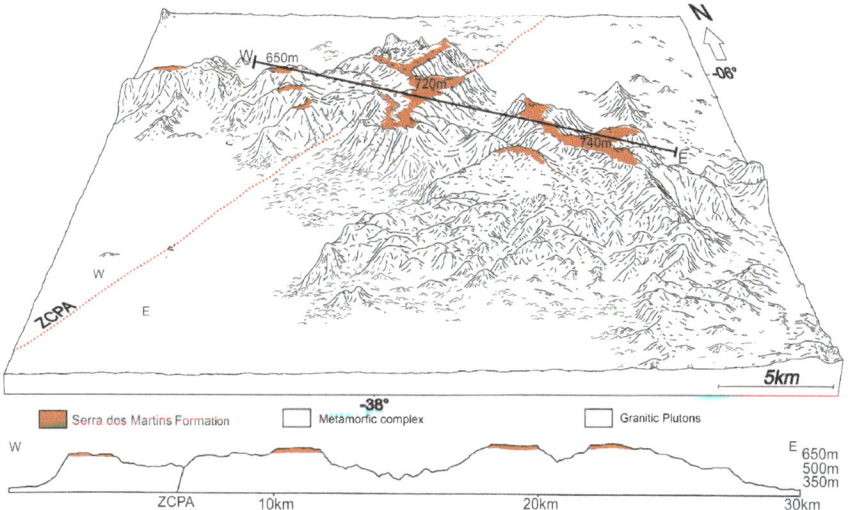

Fig. 4.3 Block diagrams of the Portalegre and Martins (*Source* Modified from Maia and Bezerra 2011)

superior Oligocene period (Morais Neto et al. 2002). Through facies analysis and the elaboration of profiles of the vertical stacking of layers and their lateral variations, Menezes and Lima Filho (1997) and Menezes (1999) revealed that the Serra dos Martins Formation is derived from a meandering river system. This system is mainly represented by channel bottom deposits, channel fill, channel overflow, and the floodplain, affected by brittle deformation (Barros 1998).

Several examples of intrusions over deformation zones can be found in the northern sector of the Borborema Province, north of the Patos Lineament (Van Schmus et al. 2011).

(a)   **(b)**

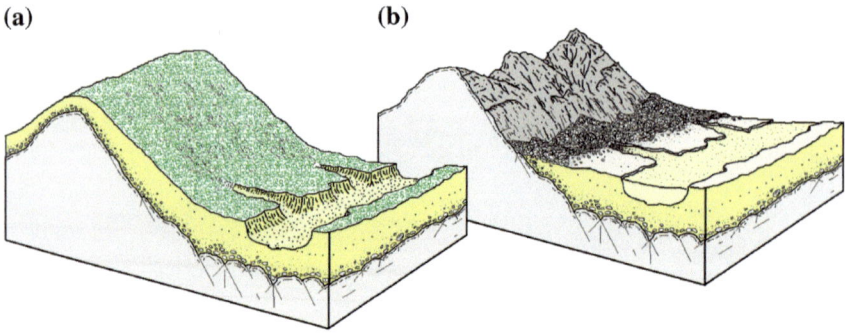

Fig. 4.4 Hypothetical erosion model in the dry season

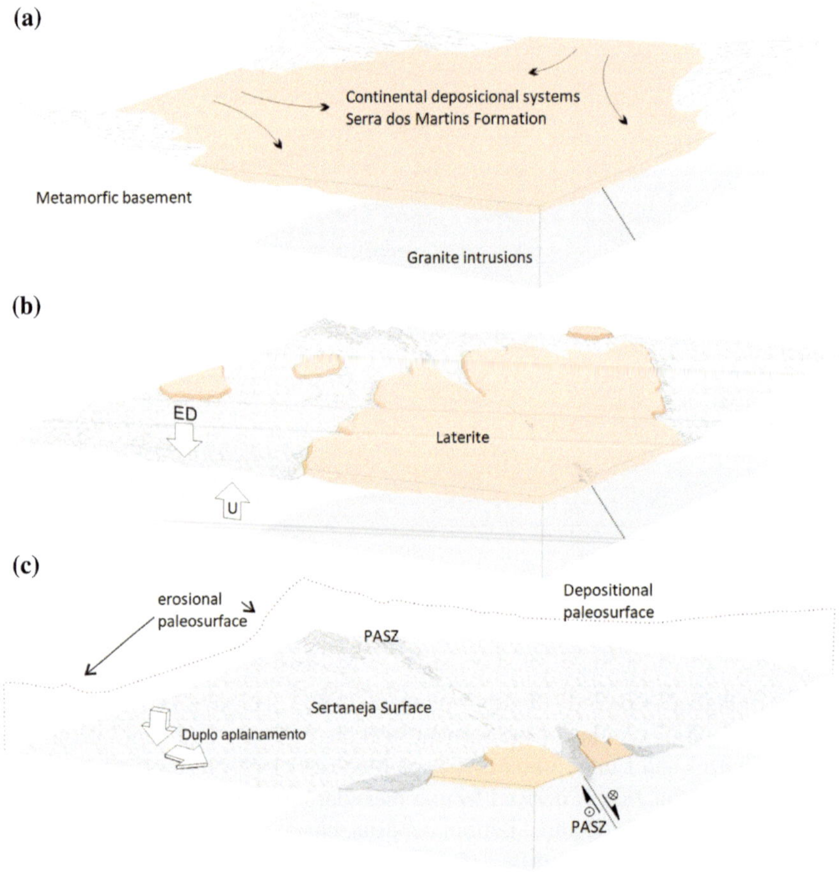

(a)

Continental deposicional systems
Serra dos Martins Formation

Metamorfic basement

Granite intrusions

(b)

ED

Laterite

U

(c)

erosional
paleosurface

Depositional
paleosurface

PASZ

Sertaneja Surface

Duplo aplainamento

PASZ

Fig. 4.5 Inversion model of the relief according to structural/denudation processes. **a** Differential erosion of the basement, **b** Fragile reactivation (compressional) along the PASZ, (U) Epeirogenesis (*Underplaiting*)

Specifically, the Portalegre and Martins Massifs occur in the area of the Itaporanga and Poço da Cruz intrusive suites. These intrusive suites correspond to a set of synorogenic granite rocks, from the Brasiliano Era (Angelim et al. 2006). Regarding the intrusive suites, the Serra dos Martins Formation (FSM), this sedimentary unit corresponds to fluvial sandstones capped by a lateritic duricrust.

The stratigraphic positioning of this sedimentary unit was established through indirect relationships with potentially chrono-correlated sediments in the Potiguar Basin (Barros 1998; Menezes 1999) or the alkali basaltic Cenozoic volcanism (Morais Neto et al. 2001). As a result, the essentially continental Serra dos Martins Formation has been considered as the proximal equivalent of the platform sediments of the Tibau-Guamaré Formations (Menezes 1999), which are a mixed platform of the regressive mega sequence of the submersed Potiguar Basin (Soares et al. 2003). However, the division of the regressive mega sequence into distinct parcels, with ages varying from the Mesocampanian to the Miocene (Person Neto et al. 2007) hinders a more precise correlation, since the continental remnants are devoid of fossils that would enable chronological relationships to be established with the sequences of the submerged portion.

Peulvast and Sales (2004) and Peulvast et al. (2008) considered the Serra do Martins Formation, the equivalent of the Paleozoic sediments of the Parnaíba Basin, suggesting that the preserved plateaus of the Borborema high are the traces of a pre-rift sequence in the northern part of the Borborema Province.

The absence of precise chronological markers in the Serra do Martins Formation has led to many interpretations of the geomorphological evolution of the region that have been mainly based on morphoclimatic processes. The data from AFTA Moraes Neto et al. (2008) suggested that the Serra do Martins Formation has deposited between 64 and 25 Ma (Paleocene–Oligocene range). On the other hand, the ages proposed by Lima (2008) using the (U-Th)/He method, from the oxides/hydroxides of iron samples from the plateaus of the Serra do Martins Formation, indicate a minimum age of 20 Ma for the deposition of the sediments of this formation. The work of Luz et al. (2015) corroborates these ages by suggesting that the post-Cretaceous denudation of the Borborema Massif would have provided the sandstones and Paleogene conglomerates where the sedimentary units of the Serra dos Martins Formation are found.

The plateaus are around 700 m in altitude, delimited by steep cliffs that occur most frequently on a NE-SW orientation. Both the massifs have flat tops and are partially covered by laterized sandstones from the Serra dos Martins Formation (FSM). From a geomorphological point of view, this formation occurs as plateaus of flat to slightly rippled relief with steep cliffs and jagged edges. Generally speaking, they form roughly two uplifted blocks individualized by incised valleys of a NE-SW direction, which corresponds to the structural direction of the Portalegre shear zone (ZCPA).

These plateaus have more abundant rainfall regimes because of the orographic factor, a fact that is reflected in the formation of developed soils, as well as plant communities, where species of flora found in the Atlantic Forest and *Cerrado* are common. These areas, whose vegetation is determined by the greater altitude of the relief are called humid enclaves and have a distinct ecological environment to the surrounding depressed areas (Souza and Olive 2006).

4.3   Structural Controls

The NE-SW-trending lineaments of the PASZ represent a superficial expression of the Brasiliano deformation with a ductile/brittle nature reactivated in the Cretaceous and the Cenozoic. These lineaments mark fault trends that have an important influence on the structural control of the drainage system, the dissection of the relief and the Quaternary deposition. Examples of this morphostructural disposition are commonly found in the erosional areas represented by the *sertanejo* depressions and the orientation of the escarpments of the crystalline massifs.

The conditioning of the erosion processes along the PASZ resulted in NE-SW-oriented valleys to the North and the South of the Portalegre Massif.

Taking the PASZ as reference for the individualization of the E and W segments of the Portalegre and Martins Massifs, it can be observed that there is a vertical dislocation of approximately 100 m of block E in relation to block W. The apatite fission track data, provided evidence that there has been uplift of block E and subsidence of block W since the Cretaceous (Nóbrega et al. 2005). This subsidence is responsible for the graben that gives origin to the Potiguar Basin to the north. For the Cenozoic period, the same authors have observed that both blocks have had a similar evolutionary history since the Neogene era, with uplift and erosion, although the denudation/cooling rate in Eastern block is more pronounced.

Another relevant point to be taken into consideration was the last episode of volcanism that was established in the Borborema Province. This is the Macau volcanism from the Paleogene to the Neogene epoch with an uprising of isotherms. This volcanic activity may have been the cause of the Neogene epeirogeny that led to the Borborema Massif as well as the various massifs that surround it (Olivaira e Medeiros 2012).

According to Oliveira and Medeiros (2012), the magma would have been trapped on the edge of the root of the lithosphere and the continental crust, creating an assimilation of subsurface magma (underplating). As a result of the difference in density, the area of the Borborema Plateau was raised due to the thrust of the underplating.

The change of the aggradational environment to an erosive one can be attributed to this uplift, which began to dissect the regional relief according to the structural plans defined by the PASZ (Maia et al. 2016).

The differential erosion that was established distinguished and enclosed the metamorphic basement of the intrusive core represented by the Itaporanga and the Poço da Cruz suites.

Still regarding the factors linked to lithological resistance, the laterized sandstone capping contributed to the geomorphological maintenance of the top, which currently demarcates the aggradational paleosurfaces of the Serra do Martins Formation. Another hypothesis raised by Peulvast and Sales (2004) and Peulvast et al.

(2008), also incorporates the effects of the rifting in the inferior Cretaceous Era. This proposal interprets the Martins and Portalegre mountains as the remains of the erosion of the southern shoulder of the Potiguar rift. Thus, the Neogene uplift associated with the underplating effect would be an extension and acceleration of the post-rift uplift that started in the Cretaceous Era (Peulvast et al. 2008). From analyzes of apatite fission tracks, Moraes Neto et al. (2009) revealed the existence of two episodes of cooling: the first aged 100–90 Ma (Lower Cretaceous) and the other aged 20–0 Ma (Neogene). The first event was associated with the uplift and denudation of the regional topography after the continental breakup. The second event was associated with climatic changes that accentuated the erosion of the shallower crustal levels causing the deposition of sediment on the Barreiras Formation along the coast.

In block W, the Serra dos Martins Formation is located between the elevations of 500 and 600 m and may reach up to 650 m. In segment E, the Serra dos Martins Formation reaches 740 m. This unevenness demonstrates that the normal component associated with the PASZ during the Cenozoic did not have the same intensity of the Cretaceous Era when the sedimentary basins were generated. The vertical displacement of approximately 100 m between blocks E and W is compatible with the rates of movement of the neotectonic faults in Northeastern Brazil verified by Bezerra et al. (2001) and Nogueira et al. (2010).

Besides the tectonic uprising, it is necessary to highlight the structural control exerted over the dissection that follows the Brasiliano deformation planes. This is due to erosion of the cataclasite and/or mylonite associated with the PASZ causing NE-SW-oriented incised valleys and escarpments in the relief.

4.4 Exhumation of the Brasiliana Shear Zones

In the shear zones, the relief is its character is commonly arranged in sequences of ridges and valleys oriented according to the trends of the lineaments (Maia and Bezerra 2014).

These trends are represented by residual crests or mylonitic foliation planes, quartzite or mica-schist crests, and aligned granitic intrusions. This is the origin of the lineaments with different resistances to physical or geochemical denudation, favoring the wear of the softer layers through differential erosion. This erosion occurs according to the foliations and the contact with intrusive bodies, which become expressed in the relief as residual crests with a structural origin.

The regional effects of the exhumation of the basement due to erosion are shown in Fig. 4.6. In the northern sector of the Borborema Province, quadrants B, C, and D are significant examples of this morphostructural configuration, arranged in residual ridges which confine the drainage of the first order and direct the dissection. In quadrant B, the area corresponding to the Glint western peripheral depression of Ibiapaba, the Transbrasiliano lineament controls the channel of the Jaibaras River and part of the Rio Acaraú. In this sector, while the Transbrasiliano Lineament is responsible for the displacement of the front of the escarpment of the Ibiapaba Glint

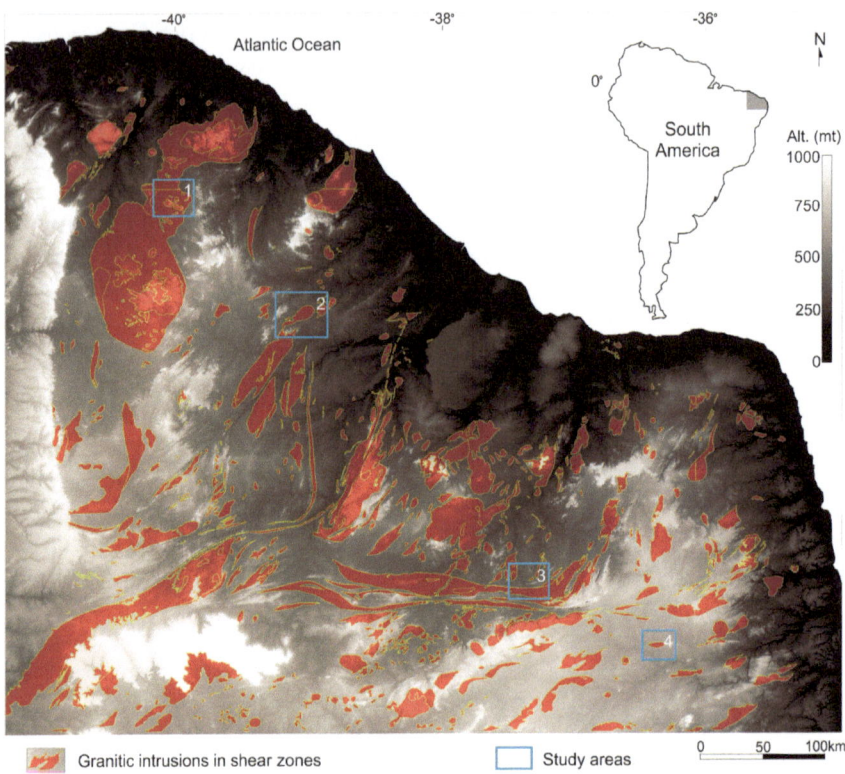

Fig. 4.6 Granitic plutons in the Northeast of Brazil

in the peripheral depression, its influence is on the fluvial dissection that resulted in a valley oriented in a NE-SW direction. In areas corresponding to the Pereiro Massif and the surrounding area of the Santana Plateau, the relief is structurally controlled by ZC Portalegre and Picuí-João Câmara. The exhumation of the ZC associated with these sectors resulted in the formation of ridges and valleys incised in a NE-SW direction that controls the flow channels of the upper reaches of the river basin of the Apodi-Mossoró River and the small basins dissecting the N and NE segments of the Borborema Massif. This morphostructural arrangement protects important tectonic indicators in its formation.

In general, control of structural relief is evident from the dissection under ductile and brittle tectonic structures. In the first case, there are mainly alignments of ridges and valleys oriented along positive and negative NE-SW- and E-W-trending lineaments that are parallel to ductile structures. In the second case, the brittle deformations control the dissection and the Quaternary sediment deposition. This occurs through the structural control of the drainage systems along neotectonic fault zones. In this regard, the largest river basins represented by the Acaraú, Jaguaribe Apodi-Mossoró, and Piranhas-Açu Rivers drain their run-off under the influence of NE-SW faults that influence the direction of valleys and therefore the fluvial aggradation. Thus, the valleys with a NE-SW direction are geomorphological expressions of Cenozoic reactivation of the Precambrian shear zones.

With the exception of the coastal plain, the other units present orientation patterns of dissection ranging from N-S, NE-SW, and E-W according to the direction of the Precambrian ductile shear zones. In the areas of massifs and depressions, the exhumation of the shear zones controls the morphological features, forming valleys and ridges of the same direction. These deformation belts individualize more dissected sectors, where structural control is less evident. These are the *sertanejo* depressions that are distributed in the inland areas as extensive flat surfaces interrupted by isolated reliefs, composed of granitic massifs that are more resistant rocks than the surrounding area. In this unit of relief, dissection processes predominate over aggradation processes generating a continuous exposure of the basement.

Thus the erosion of the ductile shear zones leads to the progressive exhumation of the Precambrian crystalline massifs. Through differential erosion, this exhumation enabled the formation of geomorphological lineaments arranged into positive and negative linear forms associated with Brasiliano foliations.

On a regional scale, the Brasiliano foliations control several morphologies, especially those associated with ductile shear zones where the relief is characterized by symmetrical crests with accentuated slopes arranged in a continuous form (Maia and Bezerra 2014).

This context is well represented in the Portalegre and Martins massifs where the exhumation of the granitic plutons (Itaporanga and Poço da Cruz) in deformation zones with a NE-SW orientation caused lineament trends that confine the drainage channels, guiding the dissection and on occasion the fluvial aggradation. This resulted in sequences of ridges and valleys that confine the flow channels that become indicators of deformation planes. This is extended to the sedimentation environment, forming embedded deposits along the structural flaws that make up the geomorphological expression of the brittle reactivation of the transcurrent shear zones.

4.5 Differential Erosion and Topographical Inversion

The option in this chapter was to use the double planation theory of Budel (1982) that occurs following the weathering and subsequent evacuation of the alteration mantle exposing the foundation and thus creating the etch form (Twidale and Vidal Romani 1994).

Budel's theory of etchplanation (1982) is widely applied in tropical areas that present seasonality and may subsidize the geomorphological interpretation of the relief in tropical environments. According to the etchplanation model, during humid periods there is a deeper alteration, whereas superficial erosion is more intense during the dry season, promoting planation and in some cases exposing the front of the alteration.

The origin of the massifs is associated with the characteristics of the substrate and the geochemical properties of the rocks that induce the increase in chemical weathering, enabling morphogenic action through fluvial processes and mass movements. These mass movements promote the lowering of the relief in less competent rocks and the most resistant remain as topographical disturbances. Thus, the alternation between the geochemical alteration of rocks and surface erosion through the variability of the climate (Bigarella 2003) makes the driest periods erosive cycles with the exposure of saprolite. This, in turn, has an irregular topography that is being exhumed through differential erosion, thus showing the nuclei of intrusive granitoid in the form of massifs.

The exposed granitic core, once submitted to the progressive differentiated deepening to the front of the alteration, associated with surface erosion, makes the sectors of the basal weathering surface that have not been altered gradually rise to the surface (Vitte 2005). Among the reasons that can generate this uplift, the lithological differentiation of the basement is noteworthy.

In the Portalegre and Martins Massifs, the capping of the SMF in the form of lateritic duricrust adds to the lithological resistance factor associated with intrusive bodies.

As demonstrated by Butt and Bristow (2013), the formation of duricrust from lateritic concretionary levels can contribute to the inversion of the relief and especially in maintaining the top. In the study area of Northeastern Brazil, the ferruginous crust that sits on the granites of the intrusive suites is a double resistance factor contributing to the further erosion of the surrounding basement. However, it is worth noting that the duricrust's contribution to the inversion process is of a lesser magnitude, about 100 m above the etch form, corresponding to the front of the exposed alteration.

The massifs in the form of plateaus are residual forms of an ancient continuous capping, partially dissected and leaving only some disconnected residual testimonies in various crystalline massifs located in the northern portion of the Borborema massif (Maia et al. 2016).

The models of the inversion of the relief based on differential erosion (Pain and Ollier 1995) or tectonic inversion (Turner and Williams 2004), although distinct, result in the inversion of the regional topography.

Thus, the structural factor associated with differential erosion of the basement leads to an inversion of the relief from the following points: (A) differentiated erosion caused by resistance to weathering of lithological units with different mineralogical compositions, (B) reversal of the stress field and the reactivation of faults or transcurrent shear zones, and (C) regional epeirogenic uplift.

In the Portalegre and Martins massifs, sedimentary remnants occurring in the top surface have been considered important evidence of the Cenozoic history of Northeastern Brazil; given that its topographical position is commonly attributed to episodes of post-Cretaceous uplift (Ab'Saber 2000). The covers are composed of continental sediments deposited on a regional planation surface developed in Precambrian crystalline rocks, commonly referred to in the literature as the *South American Surface* (King 1956) and *Borborema Surface* (Ab'Saber 1969). The sedimentary deposits form plateaus and homoclinal plateaus. The thickness of the sedimentary packages varies from a few meters to a few dozens of meters, and its upper portion is often silicified or protected by ferruginous duricrust at the top of the lateritic profiles, helping to preserve them from erosion.

As for the resulting dissection of the epeirogenic uplift, Oliveira and Medeiros (2012) used gravimetric and magnetic data to study the isostasy and causes of Cenozoic magmatism in the Borborema Province and attributed the uprising of the Borborema Massif to the Cenozoic continental magmatism generated by a convection mechanism on the edge. This process involves the generation of convection currents on a small scale due to the instability of the contact between the cold and thick continental crust and the adjacent hot mantle. According to this model, the magma was trapped on the edge of the root of the lithosphere and the continental crust, creating a subsurface assimilation of magma (underplating). As a result of the difference in density, the area of the Borborema Massif was raised due to the thrust of the underplating. This uplift with an epeirogenic cause is possibly the main modifying element of erosion base levels, changing the FSM's aggradational surface to an erosional surface, thus inducing the individualization of lithological units by differential erosion.

The crustal erosion rates in the crystalline basement defined by Thomas (1994) are: 20.8 m/ Ma (mass balance) and 22.5 m/Ma (mass balance), respectively. Extremes are 2 m/Ma (min) and 48 m/ Ma (max) for the mass balance calculations and 2 m/ Ma to 50 m/ Ma for the calculation of rates of change. This implies that the weathering rate range of the crystalline basement in different climate scenarios varies between 2 and 50 m/Ma (Ovmo 2010). For the Brazilian Northeast, more precisely the area of the Araripe Basin, the values established by Hegary et al. (2002) point to a continental denudation resulting from uplift that subtracted about 1.9 km through crustal delamination. Accordingly, considering an origin between the Paleogene and Neogene, according to AFTA data (Moraes Neto et al. 2008, Lima 2008) the exhumation values of granitic plutons that underpin the topography of the Portalegre and Martins Massifs is compatible with the dissection values presented on a general (Ovmo 2010) and local level (Hegary et al. 2002).

4.6 Granitic Inselbergs

Traditionally the works that discuss the origin and development of inselbergs, commonly do so using the concepts of down wearing and ecthplanation (Romer 2007), or by the parallel retreat of escarpments and sedimentation (King 1956). These concepts focus much of the discussion about the evolution of granitic landscapes in dry climates using a morphoclimatic emphasis. Currently, the concepts of double planation have emphasized saprolitic exhumation and exposure in surfaces from superficial alteration/removal (Tarbuck and Lutgens 2006). In both cases, inselbergs are a remnant of erosion that can provide important information on the geomorphological evolution of the land in which they occur (Matmon et al. 2013).

Fig. 4.7 Inselbergs of Quixadá—CE (Northeast of Brazil)

Most of the granite bodies found in the north of the Brazilian Northeast are associated with Brasiliano shear zones (Fig. 1.2) (Almeida e Ulbrich 2003). These shear zones are conduits through which a large volume of fluids can circulate (Trindade et al. 2008). Thus, a large number of orogenic granites are interpreted as the result of the intrusion in extensional regions associated with local and regional tectonic structures (Neves 2012). In the Borborema Province, these structures are mainly represented by shear zones with a NE-SW and E-W orientation (Vauches et al. 1995) (Fig. 4.7).

Specifically in the research area, the intrusions resulting from the rising of magma along extensional shear zones (Castro et al. 2002), mainly the Quixeramobim and Senador Pompeu shear zones (Nogueira 2004) (Fig. 1.2). Various examples of intrusions along the deformation zones can be found in the northern sector of the Province Borborema, to the north of the Patos Lineament (Fig. 1.2) (Van Schmus et al. 1995). These intrusions occurred during the Brasiliano Orogenesis (Arthaud 2007) resulting from Brasiliano/Panafricana tectonic bonding (Brito Neves et al. 2000), which was accompanied by a significant granite plutonism of 585 Ma (Fetter 2000). During the Brasiliano orogeny, numerous granitic bodies intruded the continental crust, evidencing the climax of an orogenic and magmatic event (Magini and Hackspacher 2008).

The rise of the magma occurred within the continental crust resulting in the batholiths that were exhumed by erosion and dissection of the rock wall basement below. In this regard, Silva (1989) found that the crystallization of granitic rocks in

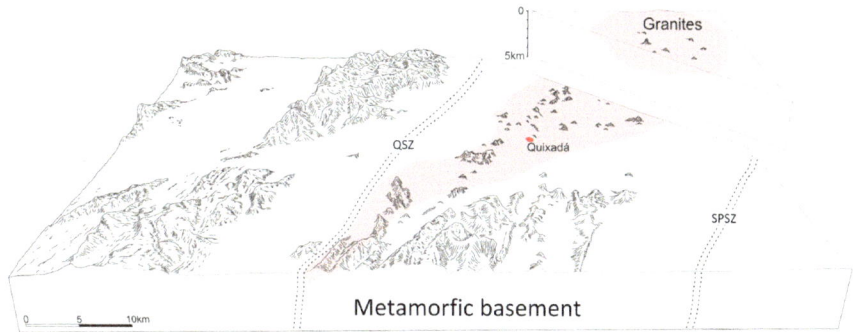

Fig. 4.8 Geological–Geomorphological Block Diagram of Quixadá and surrounding areas (QSZ—Quixeramobim Shear Zone. SPSZ—Senador Pompeu Shear Zone). Morphology of the subsurface of the granitic based batholith in de Castro et al. (2002)

Quixadá occurred under a lithostatic pressure of 6–8 kbar, the equivalent to crustal depths between 25 and 30 km. Thus, the granites have mostly a porphyritic monzonite texture nucleated by mafic enclaves (Adams, 1995). The embedded basement in the Quixadá region (CE) corresponds to the undifferentiated gneissic-magmatite complex occurring around the granite complex, occupying about 90% of the volume of the rock wall (Fig. 4.8)

The active weathering processes are subordinated to semiaridity, with a typical semiarid climate (Nimer 1989), which is characterized by the predominance of high temperatures associated with a system of sporadic rains, mostly concentrated in the first four months of the year. According to Nimer (1989), the semiarid climate is influenced by the intertropical convergence zone, with the dry season from June to January and the rainy season from February to May. The rainfall system is mainly controlled by various mechanisms including cold fronts, the position of the Intertropical Convergence Zone (ITCZ) and Upper Tropospheric Cyclonic Vortex in (UTCV) and the Trade winds (Noble 1994). The average rainfall is around 700 mm and annual temperatures are around 27 °C, with a minimum of 21 °C and a maximum of 36 °C. On average, the relative humidity throughout the year in the region is 70% and it follows the precipitation curve, with higher values observed from February to May and lower values from June to January (INMET 2014).

The vegetation on the inselbergs, the growing sites, which are micro climatically and edaphically dry, develop and sustain highly specialized vegetation (Porembski 2007). The predominant vegetation is *caatinga* scrub with a predominance of angiosperms (Gomes and Alves 2009); the soils mostly occur in Natric Planosol and Litholic Neosols associations with scattered occurrences of Vertisols in the lowest areas (Brazil 1981).

It can be verified that the pattern of lineaments has a predominantly NE-SW orientation that is related to the shear zones of Quixeramobim and Senador Pompeu. Inside the batholith, these lineaments are less significant when compared with the density of lineaments in the surrounding area enclosing the rock wall basement (gneiss-migmatite complex). This corroborates the work of Nogueira (1998) who revealed that in the central part of the batholith the effects of ductile Brasiliano deformation was less intensive, thus presenting an incipient foliation. This foliation was developed in the magmatic state and has an orientation generally parallel to the general direction of NE-SW-striking shear zones, which controlled the emplacement of the pluton (Nogueira 2004).

An analysis of the lineaments in the area of the Quixadá batholith indicated that the highest concentration of NE-SW and NW-SE linear features (structural lineaments) coincides with areas with greater space between the inselbergs. This suggests more erosion associated with these areas, controlled by structural trends that favor dissection. The lineaments are mainly represented by small topographic depressions in the form of small drainages of the first order without an apparent hydrographic pattern. Figure 1.4 shows the relation between the trend of regional lineaments and the distribution of inselbergs in the granitic batholith of the Quixadá area.

In the more fractured parts of the granites, weathering is facilitated, enabling a progressive and more intense alteration when compared with less fractured sectors. Thus, the density of the inselbergs may reflect the degree of fracturing of the rocky massif. The greater density of the fractures may lead to a greater dissection causing features related to the lineaments.

The role played by the joins resulted in differential erosion that distinguished the batholith according to the density of the fracturing. It can be seen that the higher incidence of inselbergs is related to granitic cores with lower densities of fractures, which allow their permanence as an outcrop.

Associated to this context, the paleoclimatic picture can contribute significantly to differential erosion and consequently to exhumation. This implies, among other things, that the subsurface substrate will appear on the surface because the pedogenic phases associated with morphogenetics alter and remove, respectively, the shallower crustal levels indicating deeper structures such as batholiths.

When exhumed, the inner surface of a granite body, defined by a structure that is affected by weathering processes, is much greater than the corresponding change in the external surface, usually restricted to the top of the batholith. The main result of weathering is the loss of granular rock cohesion, thus allowing the evacuation of friable debris (Vidal Romani and Temiño 2004) (Fig. 4.9).

This results in the continuous outcropping of the basement exposing irregularities in the distribution of fractures in the form of topographic smoothing. These topographic heights of the basement are the inselbergs that result from the joint action of the differential surface alteration. This alteration is related to the density of the fractures, which increase the susceptibility to chemical weathering of the granites. Thus, the differentiation of the facies associated with the structural fabric favors selective weathering resulting in different erosion. This context associated with climatic variability favors the removal of the alteration mantle resulting in inselbergs. These, in turn, reflect in their forms, the preponderance of their most significant genetic factor, be it geochemical in nature for the more soluble facies (mafic minerals) or physical in the case of inselbergs that evolve by thermoclastic weathering.

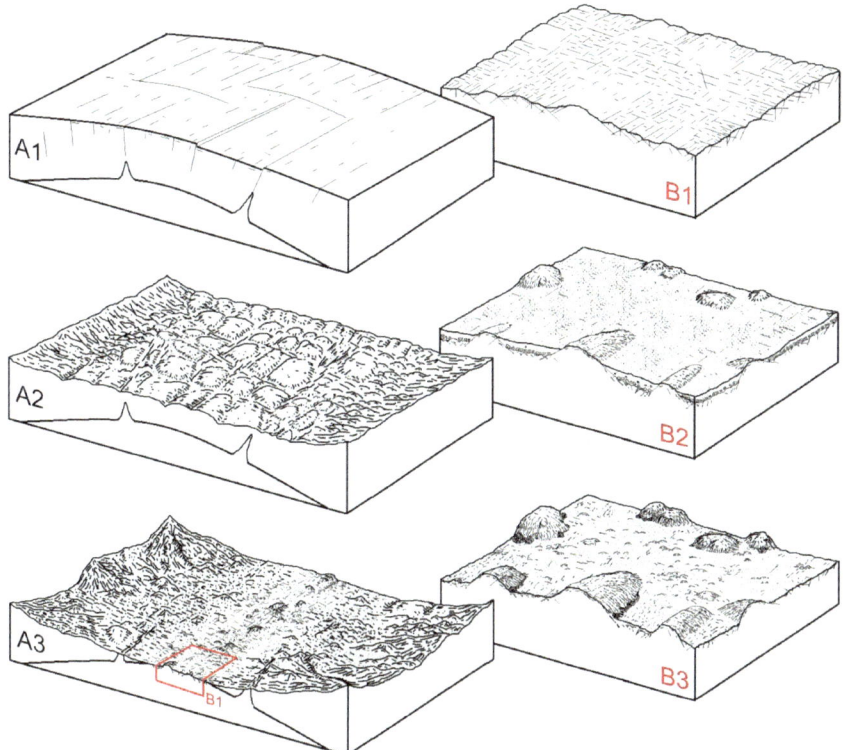

Fig. 4.9 Differential erosion and exhumation of batholiths. A1—The magma rises and exerts pressure on the overlying crust producing fracturing. A2—Fractures condition the erosive processes intensifying dissection. A3—Formation of a depression surrounded by residual reliefs that enabled the outcrop of the batholith. B1—Batholith's surface exhumed. B2—Fractures facilitate the process of surface weathering resulting in an alteration mantle. B3—In an erosional phase, the alteration mantle is removed exposing the irregularities of the basement causing inselbergs. Steps B1, B2, and B3 based on the Budel Etchplanation model (1982)

In the study area, it was possible to distinguish the occurrence of three types of inselbergs according to their morphogenetic characteristics. The identification of these inselbergs types was based on the erosional features derived from dissolution, fracturing, or the absence of both. Specifically, in regards to the formation of erosional features, it was noted that this association could be made from a lithological correlation and facies in the same granite.

The three groups of inselbergs can be organized as follows: 1—Inselbergs with features of dissolution, 2—Inselbergs with features of fracturing, 3—Inselbergs massifs (Fig. 4.10).

The dissolution inselbergs have the lowest altitudes and more features of dissolution of the weathering pits and flute marks.

Fig. 4.10 Inselbergs types: **a**—Inselbergs with features of dissolution, **b**—Inselbergs with features of fracturing, **c**—Inselbergs massifs

Fig. 4.11 Dissolution features in granitic inselberg

In these inselbergs, there are fracturing features and those of scaling is not very apparent. Usually, there are porphyritic granitic facies rich in well-developed feldspar phenocrysts with a micaceous matrix of the aphanitic type.

The dissolution features associated with these inselbergs are due to their composition. In this case, the solubility of the biotite, specifically in the mafic enclaves and feldspar phenocrysts. The lower content of the biotite associated with these cases gives greater physical cohesion and resistant to the thermoclastic weathering of the rock. These inselbergs exhibit convex morphology associated with dissolution and do not show erosional features resulting from fracturing or well-developed *scaling*. On the escarpments, a dense dissection network developed. These dissolution features are continuous or stepped-type grooves with staggered levels of weathering pits dissolution basins. The starting point for the formation of dissolution features are the mafic enclaves within the granite. In these enclaves, there is a temporary rise in the proportion of biotite relative to feldspar resulting in a more intense dissection, which engages the drainage network leading to the groove (Fig. 4.11).

The fractured inselbergs are characterized by a high density of the fractures. In these cases, the features that best characterize these inselbergs cause dismantling and the collapse of the blocks. These inselbergs are not liable to classification according to concavity or convexity patterns, exhibiting a chaotic morphology resulting mainly

Fig. 4.12 Fracture features in granitic inselberg

from thermoclastic weathering and exfoliation. On the steepest sectors, exfoliation features caused tafonis collapse.

The blocks loosened by exfoliation collapsed and caused chaotic rudaceous deposits at the base of the inselberg slopes.

The predisposition of type 2 inselbergs to fracturing is associated with the occurrence of mafic enclaves and dikes that control the propagation of thermoclastic fractures. The presence of these mafic enclaves results in the injection of mafic magma at different stages of crystallization. In this regard, Jacobs (2012) considers that if the mafic magma is injected into the crystallized rock, this magma is channeled to the early fractures in the almost solid granite and interacts with the last magmatic liquid only locally to form compound or fragmented dikes. While the very fine grain of the granites indicates rapid cooling due to the temperature contrast with the host rock (Fig. 4.12).

The third group of inselbergs exhibits massive escarpments with no apparent dissection features. There are no significant erosional features of dissolution such as flutes, Vasques and tafoni, fracturing, spheroid disjunction or scaling. These inselbergs occur mainly outside the area of the main batholith to the south and are associated with the wall rock of the basement of the gneiss-magmatic complex. These are inselbergs with steep cliffs and with convex morphology. As they occur outside the main batholith area, they are oriented along the NE-SW-striking shear zones (Fig. 4.13).

Fig. 4.13 Massive inselbergs

In the light of classic geomorphological theories, the inselbergs of semiarid areas such as Northeastern Brazil have been explained as associated with an erosive and residual origin. However, from the concept of double planation, it was found that the inselbergs, including the Borborema, are located in less fractured or more mineralogically resistant areas of the foundation (Cooke et al. 2010). Specifically, the descriptions of the relief developed in Quixadá by Torquato et al. (1989) explain the genesis and evolution of inselbergs using the pediplanation model from the lithological control.

In this regard, Goudie (2004) certified that there are relatively few cases in which inselbergs only occur by lithological control, where the base of the escarpment coincides with a lithological limit. Thus, the origin of inselbergs would not be related only to differences in the mineralogical composition between the hill and the surrounding plain, but especially by structural control by fracturing.

Thus, the contrasts in the composition or density of the fractures are sufficient to initiate differences in patterns of weathering and erosion resulting in the formation of the inselbergs (Romani Twidale and Vidal 1994).

Budel's concept of etchplanation (1982) has been applied to tropical areas that present seasonality and can underpin the geomorphological interpretation of inselbergs in the tropical environment. According to the etchplanation model, there is a deepening in the alternation during humid periods, while superficial erosion occurs with more intensity during the dry season, promoting planation and in some cases exposing the front of the alteration. Thus, understanding the geomorphological evolution of the area of inselbergs in Quixadá is to understand the mechanisms of exhumation of the batholith through erosive cycles and how these are influenced by the existing structure.

According to Vitte (2005), the origin of inselbergs is linked to the characteristics of the substrate, associated with the geochemical properties of the rocks that lead to an increased intensity of chemical weathering, enabling morphogenetic action through fluvial processes and mass movements. These mass movements promote the lowering of the relief in some less resistant rocks and the more resistant remain as outcrops in the topography. For this purpose, switching between the geochemical alteration of rocks and surface erosion by climatic variability (Bigarellla 2003) means the driest periods have erosive cycles with the exposure of saprolite. This, in turn, has an irregular topography that is being exhumed through erosion and the removal of the more finely textured grains, thus revealing the inselbergs (Maia et al. 2015).

The extreme climatic instability that occurred during the Pleistocene imprinted distinct and successive models of development on the landscape, with the alternation between wet and dry phases being the fundamental characteristic of this period (Bigarella 2003). In Northeastern Brazil, phytogeography was the main climatic impact, making the savanna vegetation expand and retract several times in the Pleistocene (Cavalcante 2005).

Different to the work of Torquato et al. (1988), which made considerations on the geomorphological behavior of facies of the basement on a regional scale, this chapter analyzes the different patterns of inselbergs within the same lithological unit (of Quixadá) that comprise a homonymous batholith.

Regarding the genesis of the physiological variability within the Quixadá batholith, Almeida et al. (1999) state that the lithological associations are the product of the magma being mixed. The magmatic enclaves found in this batholith have mineralogical and chemical features indicating that the original liquids have a basaltic composition and that these basalts, as indicated by the numerous syn-plutonic dams, were injected with a more acidic liquid at various stages of crystallization leading to chemical hybridization. In this regard, Neves (2012), states that the differentiation of basaltic magmas can produce liquids with a granitic composition.

This mineralogical variability within the same lithological unit may explain the diversity of forms that the inselbergs have, especially types one and two.

They are characterized by features that are a geomorphological response of the composition of the facies and the morphoclimatic evolution. Whereas the origin of the inselbergs is the result of the differentiation in the intensity of the deformation, be it brittle and post-intrusion in nature or ductile and therefore syn-plutonic.

The deformation events give rise to structural features of the foliation/fracturing-type favoring thermoclastic weathering, once the lineation of the minerals is a preferred area for the propagation of fractures. The fractured foundation facilitates

Fig. 4.14 Inselbergs of Quixadá

saprolitization causing a pedogenic area. In the shift from a more humid climate to a drier one, the reduction in humidity results in the expansion of the floristic deciduous systems with more spacing and less capacity to protect the alteration mantle. Erosion takes place by removing the finer fractions resulting in exposure of the saprolite. This process results in saprolite relief, sometimes with granite Tors, where partially altered loose blocks are exposed on the surface. Thus, the removal of friable debris resulting from the alternation of erosive cycles will create the inselbergs through differential erosion (Fig. 4.14).

The granitic core exposed after being submitted to the differential progressive deepening of the front of alteration, associated with surface erosion, results in the of sectors of the basal surface of weathering that have not changed being gradually raised to the surface (Vitte 2005). Among the reasons that can generate this uplift, the absence of fractures and/or differentiation facies of the basement are noteworthy.

In this regard, Torquato et al. (1988) have stated that the differentiation of the granitic facies on a regional scale can display a certain geomorphological behavior. Thus, the main geomorphological evidence of a selective behavior of the granite facies in relation to erosive action is derived from the occurrence of inselbergs, either grouped or ungrouped. In a semiarid environment where thermoclastic weathering processes predominate, some properties of granitic rocks should be considered to demonstrate their geomorphological behavior. Among the properties capable of justifying the occurrence of residual relief the follows are discernible:

1—A scarcity of biotite—increases the rock consistency resisting expansion; exfoliation planes remain united and the biotite becomes less changeable. 2—Scarcity and plagioclase and higher proportion of microcline. 3—Higher proportion of quartz. 4—Scarcity of diaclase. 5—Little porosity. 6—No saline medium.

The granite facies of Quixadá, which comprise the homonymous batholith, is among those that have most of the aforementioned conditions, justifying the highest

Fig. 4.15 Saprolitic outcrop—Lajedo de Pai Mateus—Paraíba—BR

concentration of inselbergs. The other facies may eventually require the presence of residual reliefs, but with greater spacing between them (Torquato et al. 1988).

As faciological variability is an important element for forming inselbergs, it is understood that in cases where this variability is not high, the potential for inselberg formation will be limited. In this regard, Goudie (2004) pointed out that there are relatively few cases in which inselbergs occur by lithological control alone, where the basement of the escarpment coincides with a lithological limit. Thus, the origin of inselbergs is not only related to differences in mineralogical composition between the hill and the surrounding plain but also by a structural control through fracturing.

In the most fractured sectors, weathering is facilitated, enabling a more intense progressive alteration when compared to less fractured sectors. By this means, inselberg density may reflect the degree of fracturing of the rock massif. The highest density of fractures may lead to a greater dissection while the lowest will form inselbergs (Maia et al. 2015).

In the scale of the outcrop, the exhumation of the basement leads to exposure of modified rock on the surface. This exposure occurs because of erosion of the mantle of alteration that can occur in the passage from a wet to a dry climate and originates granite Tor type reliefs (Fig. 4.15).

The exhumation leads to the alleviation of lithostatic pressure leading to the fracturing of the outcrop along planes parallel to the exposed surface. This leads to the formation of exfoliation structures by spheroidal weathering (Fig. 4.16).

Fig. 4.16 Different examples of granite exfoliation and boulders genesis

References

Ab Sáber AN (1969) Participação das superfícies aplainadas nas paisagens do Nordeste Brasileiro. IGEOG-USP, Bol. Geomorfologia, SP, n 19, 38 p

AB Sáber AN, Bigarella JJ (1961) Considerações sobre a geomorfogênense da Serra do Mar. Boletim Paranaense de Geografia n. 4/5, p 94–110

Almeida FFM, Brito Neves BB, Carneiro CDR (2000) The origin and evolution of the South American Platform. Earth Sci Rev 50:77–111

Andrade GO, Lins R (1965) Introdução à morfoclimatologia do Nordeste do Brasil. Arquivos do Instituto de Ciências da Terra, Recife 3–4:11–28

Angelim LD, Medeiros VC, Nesi JR (2006) Programa Geologia do Brasil—PGB. Projeto Mapa Geológico e de Recursos Minerais do Estado do Rio Grande do Norte. Mapa Geológico do Estado do Rio Grande do Norte. Escala 1:500.000. Recife: CPRM/FAPERN

Arthaud MH (2007) Evolução neoproterozóica do grupo Ceará (domínio Ceará central, NE Brasil): da sedimentação à colisão continental brasiliana. 170 f. Tese (Doutorado em Geociências)-Universidade de Brasília, Brasília

Barros SDS (1998) Aspectos Morfo-Tectônicos nos Platôs de Portalegre, Martins e Santana/RN Dissertação de Mestrado PPGG—UFRN

Bezerra FHR, Vita-Finzi C (2000) How active is a passive margin? Paleoseism Northeastern Brasil. Geol 28:591–594

Bezerra FHR, Amaro VE, Vitafinzi C, Saadi A (2001) Pliocene-Quaternary fault control of sedimentation and coastal plain morphology in NE Brazil. J South Am Earth Sci 14:61–75

Bezerra FHR, Neves BBB, Correa ACB, Barreto AMF, Suguio K (2008) Late Pleistocene tectonic-geomorphological development within a passive margin—the Cariatá trough, northeastern Brazil. Geomorphology 01:555–582

Bezerra FH, Rossetti DF, Oliveira RG, Medeiros WE, Neves BB, Balsamo F, Nogueira FC, Dantas EL, Andrades Filho C, Góes AM (2014) Neotectonic reactivation of shear zones and implications for faulting style and geometry in the continental margin of NE Brazil. Tectonophysics (Amsterdam) 614:78–90

Bigarella JJ (1994) Estrutura e Origem das Paisagens Tropicais, vol 1. Florianópolis: Ed. UFSC

Bigarella JJ (2003) Estrutura e Origem das Paisagens Tropicais, vol 3. Florianópolis: Ed. UFSC

Brito Neves BB (1999) América do Sul: quatro fusões, quatro fissões e o processo acrecionário andino. Bahia. VII Simpósio Nacional de Estudos Tectônicos, SBG, pp 11–13

Brito Neves et al (2014) The Brasiliano collage in South America: a review. Brazilian J Geol 44(3), São Paulo July/Sept

Brito Neves BB, Santos EJ, Van Schmus WR (2000) Tectonic history of the Borborema Province, northeastern Brazil. In: Cordani UG, Milani EJ, Thomaz Filho A, Campos DA (eds.) Tectonic evolution of South America. Rio de Janeiro, 31 International Geological Congress, pp 151–182

Correa ACB, Tavres BAC, Monteiro KA, Cavalcanti LCS, Lira DR (2010) Megageomorfologia e Morfoestrutura do Planalto da Borborema. Revista do Instituto Geológico, São Paulo

Corsini M, Lambert DE Figueiredo L, Caby R, Féraud G, Ruffet G, Vauches A (1998) Thermal history of the Pan/African—Brasiliano Borborema province of the northeast Brazil, deduced from $^{40}AR/^{39}AR$ analysis. Tectonophysics 285

De Castro DL, Bezerra FHR, Sousa MOL, Fuck RA (2012) Influence of Neoproterozoic tectonic fabric on the origin of the Potiguar Basin, northeastern Brazil and its links with West Africa based on gravity and magnetic data. J Geodynam 54:29–42

Furrier M, Araújo ME, Meneses LF (2006) Geormorfologia e Tectônica da Formação Barreiras no Estado da Paraíba. Geologia USP, Sér. Cient., São Paulo, 6(2):61–70

Gurgel SPP, Bezerra FHR, Corrêa ACB, Marques FO, Maia RP (2013) Cenozoic uplift and erosion of structural landforms in NE Brazil. Geomorphology (Amsterdam) 186:68, 68–84

Jordan C, Schott R (2005) Application of wavelet analysis to the study of spatial pattern of morpho-tectonic lineaments in digital terrain models: a case of study. Remote Sensing Environ 94:31 38

King LC (1956) A geomorfologia do Brasil oriental. Revista Brasileira de Geografia, Ano XVIII no 2

Lima CCU (2010) Evidências da Ação Tectônica nos Sedimentos da Formação Barreiras Presentes do Litoral de Sergipe ao Norte da Bahia. Revista de Geografia (Recife), Esp. 01:148–160

Mabesoone JM, Castro C (1975) Desenvolvimento geomorfológico do Nordeste Brasileiro. Boletim do Núcleo Nordeste da Sociedade Brasileira de Geologia. 3:3–5

Neves BB, Ochoa FL (2012) Contribution to the stratigraphy of the onshore Paraíba Basin, Brazil. Anais da Academia Brasileira de Ciências (Impresso) 84:313–334

Maia RP, Bezerra FH, Sales VC (2010) Geomorfologia do Nordeste: Concepções clássicas e atuais acerca das superfícies de aplainamento. Revista de Geografia (Recife) 27:6–19

Maia RP (2012) Geomorfologia e Neotectônica no Vale do Rio Apodi-Mossoró NE/Brasil. Tese de Doutorado. Programa de Pós-graduação em Geodinâmica e Geofísica. PPGG – UFRN. Natal, RN

Matos RMD (1992) The Northeast Brazilian rift system. Tectonics 11(4):766–791

Matos RMD (2000) Tectonic evolution of the equatorial South Atlantic. AGU Geophys Monogr. In: Mohriak WU, Talwani M (eds.) Atlantic rifts and continental margins, vol 115, pp 331–354

Menezes MRF (1999) Estudos sedimentológicos e contexto estrutural da Formação Serra dos Martins. Dissertação de Mestrado PPGG-UFRN

Moura-Lima EN, Sousa MO, Bezerra FH, de Aquino MR, Vieira MM, Lima-Filho FP, da Fonseca VP, do Amaral RF (2010) Sedimentação e deformação Tectônica cenozóicas na porção central da Bacia Potiguar. Geologia USP. Série Científica, 10

Nascimento RSC (1998) Petrologia dos Granitóides Brasilianos associados a zonas de cisalhamento Remígio-Pocinhos (PB). Programa de Pós-graduação em Geodinâmica e Geofísica. PPGG-UFRN, Natal, RN, Dissertação de Mestrado

Neves SP (1991) A zona de cisalhamento tauá, ceará: sentido e estimativa do deslocamento, evolução estrutural e granitogênese associada. Revista Brasil.eira de Geociências 21, vol 2

Nimer E (1989) Climatologia do Brasil. 421 p IBGE, Rio de Janeiro, RJ, Brasil

Nóbrega MA, Sá JM, Bezerra FH, Neto JH, Iunes PJ, Guedes S, Saenz CT, Hackspacher PC, Lima-Filho FP (2005) The use of apatite fission track thermochronology to constrain fault movements and sedimentary basin evolution in northeastern Brazil. Radiat Measurem Amsterdam, 39:627–633

Nogueira FC, Bezerra FHR, Fuck RA (2010) Quaternary fault kinematics and chronology in intraplate northeastern Brazil. J Geodyn 49:79–91

O'Leary DW, Friedman JD, Pohn HA (1976) Lineament, linear, lineation: some proposed new standards for old terms. Geolog Soc Am Bull 87:1463–1469

Oliveira RG, de Medeiros WE (2012) Evidences of buried loads in the base of the crust of Borborema Plateau (NE Brazil) from Bouguer admittance estimates. J South Am Earth Sci 37

Rossetti DF, Bezerra FH, Góes AM, Valeriano MM, Andrades-Filho CO, Mittani JC, Tatumi SH, Brito-Neves BB (2011) Late quaternary sedimentation in the Paraíba Basin, Northeastern Brazil: landform, sea level and tectonics in Eastern South America passive margin. Palaeogeogr Palaeoclimatol Palaeoecol 191–204

Sabins Jr., FF (1996) Remote sensing: principles and interpretations. Freeman and Company, 494 p

Soares UM, Rossetti EL, Cassab RCT (2003) Bacia Potiguar. Fundação Paleontológica Phoenix. Bacias Sedimentares Brasileiras, Ano 5, n 55

Trindade IV, Martins Sá J, Macedo, MHF (2008) Comportamento de elementos químicos em rochas mineralizadas em ouro na Faixa Seridó, Província Borborema. Revista Brasileira de Geociências 38(2), SP

Turner JP, Williams GA (2004) Sedimentary basin inversion and intra-plate shortening. Earth Sci Rev 65:277–304

Van Schmus WR, de Brito Neves BB, Hackspacher P, Babinski M (1995) U/Pb and Sm/Nd geochronologic studies of the eastern Borborema Province, Northeastern Brazil: initial conclusions. J South Am Earth Sci 8(3/4):267–288

Van Schmus WR, Kozuch M, Brito Neves BB (2011) Precambrian history of the Zona Transversal of the Borborema Province: insights from Sm-Nd and U-Pb geochronology. J S Am Earth Sci 31:227–252

Vauchez A, Neves S, Caby R, Corsini M, Egydio-Silva M, Arthaud M, Amaro V (1995) The Borborema shear zone system, NE Brazil. J South Am Earth Sci 8

Chapter 5
The Morphostructural Evolution of Cretaceous Basins

Abstract Most studies on the geomorphology of sedimentary basins deal with the evolution of erosion features and do not analyze the origin of morphostructures. These structures correspond to fault reactivation and other types of tectonic deformation. Examples of these morphostructures can be found in the Potiguar Basin, Northeastern Brazil, where they are controlled by processes of post-rift reactivation. These processes comprise fault reactivation and folding under a compressional stress regime and influenced deformation structures, erosion processes, and control the drainage system. In this context, the Potiguar Basin, located in the equatorial margin of Brazil, was affected by deformation events in the Cenozoic. These events generated morphostructures, which have influenced the geomorphological evolution of the basin. The Cenozoic stress field deformed the top of the post-rift sedimentary units and formed fold-like dome structures, which control the drainage system, erosion, and Quaternary deposition of sediments. The most important landform units in the basin (Mel and Mossoró highs and Mossoró and Açu River valleys) exhibit evidence of structural processes in their genesis. We conclude that the Cenozoic stress fields influenced fault reactivation, landforms, height of sedimentary Neogene deposits, and the drainage system.

Keywords Neotectonics · Basin inversion · Relief

In Northeastern Brazil the origin and evolution of Cretaceous Basins are directly related to the tectonic reactivation of the Brasiliano shear zones. This reactivation was responsible at first by forming the accommodation space for sediments deposition. Next the brittle Cenozoic reactivation of shear zones under a compressive regime generated the inversion of basins from the deformation of their rift and post-rift sections (Maia 2012). A typical example of this process can be seen in the Potiguar Basin where the reversal of stress fields has led to the reactivation of normal faults as strike-slip structures deforming the surface and causing structural landforms. In this context this chapter discusses aspects concerning the inversion processes of basins focusing on the case of the Potiguar Basin using geological geomorphological and geochronological data.

© The Author(s), under exclusive license to Springer Nature Switzerland AG 2020
R. Maia and F. Bezerra, *Structural Geomorphology in Northeastern Brazil*,
SpringerBriefs in Latin American Studies,
https://doi.org/10.1007/978-3-030-13311-5_5

5.1 Neotectonic Stress Field and Basins Inversion

The recognition of structural reliefs resulting from compression in basins initially dominated by extensional faults has had wide recognition in the literature, from the works of Williams et al. (1989), Underhill and Patterson (1998), Cipollari et al. (1999), Ascione and Roman (1999), Muñoz et al. (2002), Turner and Williams (2004), Zanchi et al. (2006) and Dore et al. (2008).

Structural reliefs resulting from compression in sedimentary basins occur primarily in the form of tectonic deformation structures in the rift or post-rift section, which resulted from the inversion of normal stress regime to strike-slip or reverse stress regime in a sedimentary basin (Turner and Williams 2004).

The inversion of a sedimentary basin occurs when normal faults are reactivated in a compressional environment stress regime, promoting increased erosion in the basin and conditioning the adaptation of drainage to the new relief conditions (Turner and Williams 2004).

In Northeastern Brazil, sedimentary basins record important episodes of their morphotectonic evolution in the relief. The marks of this evolution are imprinted in different ways, the faults, joints, folds, and other tectonic deformations have the utmost importance for the morphostructural evolution (Maia 2012). Arranged mainly in the form of plateaus, individualized by the *sertanejo* depressions, the Cretaceous sedimentary basins are affected by Cenozoic tectonic compression, displaying a vast collection of structures and deformational processes (Marques et al. 2015).

Most studies of these basins have concentrated on the rift phase. In various regions, the post rift crustal movements, especially the post-Oligocene, are poorly investigated, or not at all, propitiating the false idea that these basins represent stable areas after the Cretaceous rifting that occurred during the South Atlantic opening (Bezerra et al. 2008). However, recent studies show otherwise.

During the Cenozoic era, the reactivation of major fault systems in the Potiguar Basin resulting in folding with large wave length and axes oriented preferentially in the N-S direction, resulting from E-W compressive stresses (Cremonini and Kraner 1995).

The effects of these stresses on the morphology, drainage, and depositional environments have been analyzed by Bezerra and Vita-Finzi (2000), Bezerra et al. (2001, 2008), Nogueira et al. (2010), Moura Lima et al. (2010), Rossetti et al. (2011), Maia (2012) and Marques et al. (2014). These studies show that the relationship between the Cenozoic tectonic and Neogenic and Quaternary deposits in the Potiguar Basin is responsible for the conformity in the arrangement of the valleys, cliffs, and neotectonic faults. Several alignments of valleys and depressed areas are orientated according to the Precambrian and Cretaceous basement faults, which may represent a recent reactivation of these lines of weakness (Moura Lima et al. 2010).

Understanding the effects on the geomorphology resulting from the reactivation of neotectonic faults in sedimentary basins is still a little discussed topic, especially in the sedimentary basins of the Brazilian Atlantic shore. Thus, this chapter addresses geomorphological aspects, combining them with data on the geology, outcrops, and lithological and bibliographic profiles of wells regarding the reactivation of faults and the effects the *post-campanian* tectonic stresses in the central portion of the Potiguar Basin and the Araripe Basin. It will be shown that a significant proportion of the relief is neotectonic basins and comes from the inversion of the tectonic stress field during the Cenozoic era.

5.2 Geological and Geomorphological Characterization of the Potiguar Basin

Northeastern Brazil exhibits a series of interior and coastal basins of Neocomian age (Matos 2000; de Castro et al. 2012). Their structural framework consists of a rift phase associated with the system of faults resulting from the extensional efforts that culminated in the South America–Africa breakup in the Mesozoic. For example, the Potiguar rift presents the reactivation of NE-SW-striking shear zones in the crystalline basement during the early Cretaceous (Nobrega et al. 2005; de Castro et al. 2012). Its stratigraphic record includes three super sequences: rift, post-rift, and drift (Matos 1992). Another classification indicates that the sedimentary infilling of the Potiguar Basin is closely related to the different stages of its tectonic evolution: two phases of rifting (Rift I and Rift II), whose stratigraphic record corresponds to the set of continental sequences that make up the Super Sequence Rift; a phase here called post-rift, corresponding to the continental and marine Super Sequence, and the thermal phase, consisting of sets of transgressive and regressive marine sequences that make up the Super Sequence Drift (Pessoa Neto et al. 2007).

Currently, from the point of view of the relief, the emerged portion of the Potiguar Basin is a low cuestiform plateau.

In the front, the steep cornice is supported by carbonate rocks of the Jandaíra Formation with a somital of approximately 140 m. The Jandaíra Formation consists of carbonate rocks of Turonian to Campanian Age (94–89 Ma) and is the main outcropping unit of the Potiguar Basin. Regarding this formation, the solubility of carbonate rocks associated with the escarpment zone does not favor the formation of talus deposits at the interface between the front and the surrounding peripheral depression. The front forms a ramp with very variable slope shaped in the Açu Formation sandstones. This sector presents an incipient dissection in the form of circum denudation carried out by channels of the first and second order of anaclinal and ortoclinal types toward the *Sertanejo* Depression limiting the area of the Precambrian basement and the sedimentary basin (Maia et al. 2012).

The Potiguar Basin is dissected in its central portion by the Mossoro and Assu Rivers, which are cataclinal channels that break through the front of the cuesta developing their courses on the carbonate outcrops of the Jandaíra Formation. These two rivers have an interfluvial dome (Serra do Mel) that reaches 270 m above sea level. This dome is characterized by an NE-SW-oriented topography and is situated

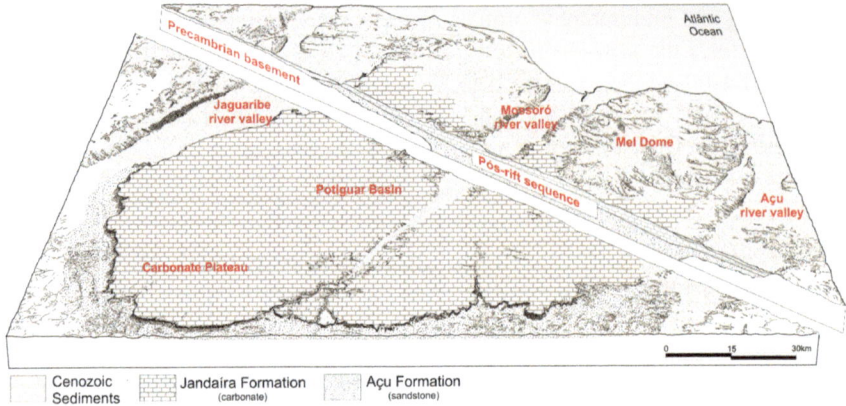

Fig. 5.1 Block diagram of the Potiguar Basin. Oblique cut showing the post-rift sequence (modified from Maia and Bezerra 2014)

in the central portion of the Potiguar Basin. Its somital is the maximum altitude of the basin and its uplift has conditioned the current drainage system (Maia 2012).

To the north, the sector closest to the shore, the relief is characterized by altimetric amplitudes ranging between 0 and 200 m above sea level, with steep coastal cliffs modeled in the Barreiras Formation. In the valleys, seasonal flood plains have an altimetric value between 0 and 4 m starting from the mouth up to 26 km inland for the Mossoró River and 8 km for the Açu River. The altimetric amplitude varies between 0 and 1 m in the limit of the river plain and the tidal river plains and 35 m in the sector where the rivers enter into the Potiguar Basin (Maia 2012) (Fig. 5.1).

According to Bezerra and Vita Finzi (2000), Bezerra et al. (2011) and Reis et al. (2013), from an evolutionary point of view from the analysis of field data, focal mechanism, and profiles of images and breakout, the Cenozoic registers two stress fields: one that occurred in the Paleogene and the other from the Neogene to the Quaternary. The first field was characterized by N-S-oriented compressive stresses and E-W-oriented oriented extension. The second field is characterized by NW-SE-oriented compression and NE-SW-oriented distension extension in the central and western part of the Potiguar Basin. This latter field affects all of the lithostratigraphic units and is the only stress field whose effects are observed in the Miocene Barreiras Formation and in Quaternary sediments. These stress fields resulted in a series of morphologies associated with the compressive context that controlled the processes of geomorphological evolution, as discussed below.

5.3 Extensional Basins and Compressional Tectonics

In the Brazilian Northeast, the Cenozoic stress fields reactivated older fault systems in a compressional regime. The reactivation of these fault systems has generated deformations in the rift and post-rift sections of the basins located on the Northeastern Atlantic coast. Specifically, in the Potiguar Basin, its altitudes were an important geomorphological expression of the deformation of the post-rift section. This deformation may be verified from the data of outcrops and wells showing that the boundaries between rock units are deformed. In the central part of the Potiguar Basin, the occurrence of a topography arranged as a domical antiform suggests the existence of an inversion feature in the post-rift section, represented mainly by the Serra do Mel (Fig. 5.1, Area B1). On the Serra do Mel, according to data on wells, the Cretaceous–Neogene boundary is located at altitudes ranging from 70 to −70 m; for the Serra de Mossoró (Fig. 5.1, Area C) field data shows that these dimensions exceed 200 m. This contact is defined by the top of the Jandaíra Formation, which is below sea level in the coastal zone.

The maximum altitude of the peak of the Serra do Mel coincides with the top elevations of the Serra de Mossoró whose summit reaches 270 m. This configuration indicates that the top of the post-rift section on Serra do Mel may be higher than evident in wells in higher locations where the altitude exceeds 270 m. These variations in the altitude of the top of the post-rift section (top of the Jandaíra Formation) may have their origin associated with tectonic processes, which have possibly influenced the geomorphological evolution of the Potiguar Basin.

Unlike the Serra do Mel, which is capped by Neogenic and Quaternary deposits, in the Serra de Mossoró, it is possible to find outcrops from the Jandíra Formation over 200 m above sea level. This is a residual relief sustained by an arenitic top from the Barreiras Formation representing a factor of resistance to the surrounding erosion that favors the maintenance of the topography. This arenitic top is silicified (Fig. 5.2), a fact that corroborates the hypothesis that the Serra de Mossoró is the result of the reactivation of Neogenic faults, as the silicification process requires the existence of faults that can conduct silicone fluids, as their genesis is associated with *post-Campanian* tectonic reactivations.

The Mossoro and Mel mountains confine the channels of the Mossoro and Açu Rivers whose geometry is demarcated by the morphostructural organization of the area defined by the lineaments of trends with an NE-SW and NW-SE direction. This configuration is shown in the morphology of the regional drainage that is controlled by two NE-SW direction elevations with a radial pattern (Fig. 4.3). In the area, the centrifugal radial drainage disperses the channels to both sides of the Serra do Mel, toward the back of the valleys of the Mossoro and Açu Rivers. A network of parallel ravines is also imprinted in the relief, with an orientation perpendicular to the river valleys, causing linear incisions and colluvial deposition.

The spatial distribution of the abandoned terraces of the Açu River is more evident west of the river, conferring the predominance of Quaternary alluvium in this portion. The provision of these sediments in the form of channels suggests that the migration

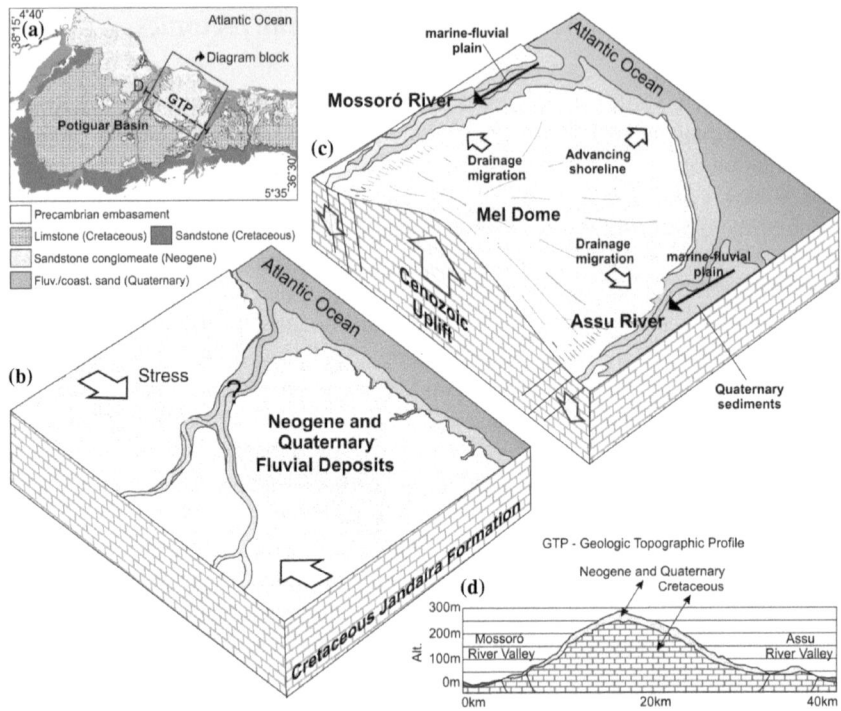

Fig. 5.2 Training model of NE-SW antiforms in the central part of the Potiguar Basin. **a**—Location of the area. **b**—Deformation model from reactivation of faults in the second compressional environment according to Jackson et al. (1996). **c** and **d**—Block diagram of interfluves and drainage organization in the valleys of the Mossoró and Açu Rivers, respectively (*Source* Maia and Bezerra 2014)

of the channels of the Mossoró and Açu Rivers is directly influenced by uplift and the formation of a structural antiform with an NE-SW axis in the central part of the basin, since the older deposits of these channels are located preferentially in directions opposite to the direction of current migration. These deposits start from lowered elevations, from the seasonal flood plains of the Mossoró and Açu Rivers to increasingly higher dimensions toward the Serra do Mel ridge. The presence of Quaternary alluvial deposits at elevations ranging from the current base level to over 200 m in the Serra do Mel is an important chronostratigraphic marker for neotectonics. In this regard, the formation of a domical antiform in the center of the basin originated a prominent interfluve that scattered the channels of the Mossoró and Açu Rivers that migrated in opposite directions. The area of uplift raised the old alluvium to successively higher altitudes as they move away from the area of seasonal floodplains of their respective rivers.

In addition to the formation of a domical antiform which indicates the main drainage of the Mossoró and Açu Rivers, the interfluves formation, also with an NE-SW direction, occurs at the base on both sides of the Serra do Mel. These interfluves have similar characteristics, as they are associated with the same tectonic context. On the formation of these interfluves in tectonically active environments, Jackson et al. (1996) proposed that the drainage, in these cases can demonstrate the effects of surface deformation caused by the reactivation of faults in compressional conditions (Fig. 5.2).

This compressional regime, active since the Miocene, is characterized by principal stresses of a (σ1) E-W and NW-SE direction. These stresses are compatible with the development of NE-SW antiforms that in this area are expressed in the form of a central dome in an NE-SW direction (Serra do Mel) and interfluves with their lateral limits in the same direction. These interfluves control the channels of the third and fourth order and may result in less significant deformations, adapting to the model suggested by Jackson et al. (1996) for compressional basins. In this model, the reactivation of normal faults in compressive regimes is responsible for the formation of topographic upheavals, whether in the form of domes or as interfluves. Thus, it is important to note that the formation of structural reliefs where there is fluvial sedimentation will be subject to topographic depressions. The high places formed from the compressive stresses can increase to higher altitudes the ancient fluvial deposits that become important indicators of mechanisms of force in evolution and the migration of channels in response to the uplift.

In the coastal part north of Serra do Mel, the geomorphological expression of their uplift is expressed in the physical geography of the coastline that advances over the surrounding area, about 10 km from the ocean in an N-NE direction. In this sector, the cliffs in the Barreiras Formation reach 100 m in altitude less than 500 m from the beach. This altimetric elevation, for the cliffs of the Barreiras Formation on the coastline, has no similar example in the northeastern equatorial Atlantic coast (Fig. 4.2). Currently, one can find various erosional features of the tafonis and alveoli type in these cliffs. These features are associated with the corrosion done by the granular disintegration resulting from wind abrasion (Figs. 5.3 and 5.4).

At some point, the elevation of 100 m is reached less than 500 m from the coastline. From this elevation, a less steep ramp begins in the direction of the Serra do Mel, reaching 200 m in altitude approximately 12 km from the coastline. At that distance from the coastline, toward the Northern Atlantic seaboard of the Northeast, the average altitudes vary between 30 and 60 m.

The NW-SE lineaments play an important role in the control of the physiography of the coastline arranged in two segments that also have an NW-SE direction. In this sector, the cliffs of the Barreiras Formation are crossed by small valleys incised in the same direction. Thus, the physiography of the shoreline associated with the arrangement of the cliffs and the direction of the valleys that dissect them are the geomorphological expression of the NW-SE lineaments of the coastal portion of the Serra do Mel.

This configuration between tectonics, structural lineaments, and geomorphology were also described in the Basin of Paraíba by Rossetti et al. (2011), where topographical breaks are associated with escarpment faults, which normally are dissected and limit the river valleys. Among these, uplifted horsts tables are formed, matching the cliffs of up to 50 m along the coast (Rossetti et al. 2013).

Fig. 5.3 Cliff in Barreiras north of Serra do Mel (*Source* Maia and Bezerra 2014)

Fig. 5.4 Tafoni and alveoli in cliffs of the Barreiras Formation

Thus, the central part of the Potiguar Basin can be defined in terms of relief as two NE-SW antiforms which individualize valleys in the same direction. Among the antiforms, represented by the Serras do Mossoró and do Mel and the valleys, the topographies reach altitudes of 270 m. These dimensions are even greater when one considers the peak of the post-rift section, as data from wells indicates the occurrence of depths of up to −70 and data on outcrops, which can reach up to 250 m. These differences in altitude in post-rift section show the feature of basin inversion resulting from the reactivation of faults from the change in stress field.

5.4 Sedimentary Tectonics and Basin Inversion

According to Lugt et al. (2003), the inversion of a sedimentary basin takes place when extensional faults are inverted by reverse faults when there is a change in the direction of the forces. The majority of the research on tectonic inversion concentrated on cases where the direction of extension and compression are similar and perpendicular to the orientation of the extensional faults (Quintana et al. 2006). One possible effect of a tectonic inversion is due to the development of domical antiforms derived from the deformation of the rift and/or post-rift section of a basin (Williams et al. 1989 and Dore et al. 2008) (Fig. 5.5).

From the proposal of deformational models resulting from the reactivation of extensional faults in a compressional regime on a laboratory scale (sandbox), it was verified that the development of compressional structures such as inversion basins are controlled by preexisting faults and that these faults that originate from an extensional scheme are invariably reactivated during compression (Ventisette et al. 2006). These types of structures were described using seismic stratigraphy in the Salar de Atacama Basin, in Northeast Chile by Muñoz et al. (2002) in the form of asymmetric anticlines resulting from the Cretaceous compression.

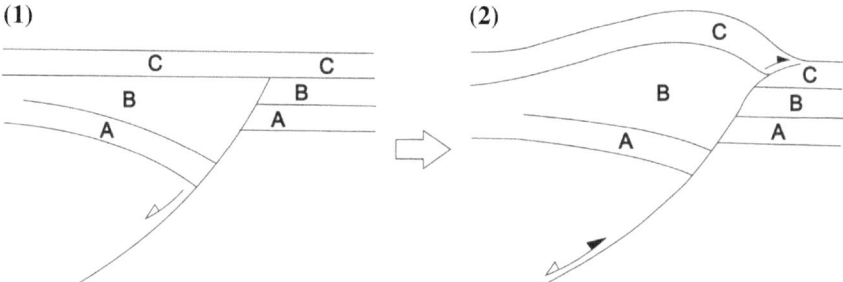

Fig. 5.5 Inversion model in Williams et al. (1989). (1) Represents the pre-inversion stage and (2) the reactivation of the normal faults of distensive regime in an inverse compressive regime. **a**—Pre-rift sequence, **b**—Syn-rift Sequence, **c**—Post-rift sequence (*Source* Maia and Bezerra 2014)

In the Potiguar Basin, the sequence of compressive events began after the extensional formational stage of the basin in the Cretaceous. In this case, the tectonic inversion is associated with compressive components, in an NNW-SSE orientation, due to a regional dextral shear in a WNW direction, related to the early rifting of the equatorial margin (Borges 1993).

In the central portion of the Potiguar Basin, recent studies have revealed features indicative of reactivation and tectonic reversals of the main normal rift faults, ranging from the basement to the post-rift section, including affecting Neogenic coverings of the Barreiras Formation (Person Neto et al. 2008). In this sector, the geomorphological evolution is related to a possible inversion of the basin resulting from current compressive stress effects ($\sigma1$) in an NW-SE direction in the Potiguar Basin since the Miocene (Souza and Bezerra 2005). This inversion is characterized by a dome (Serra do Mel), elongated in the NE-SW direction and limited in the NE by the coastal belt, in the SW by fault zones of an NW-SE orientation, in the SE by the Açu River valley, and in the NW, W, and SW through the valley of the Mossoró River (Fig. 5.6).

Studies involving seismic mapping and structural analysis of areas located over the structural trend of the main faults of the rift phase in the Potiguar Basin revealed structural and stratigraphic features that have allowed three pulses of tectonic inversion with distinct stress fields to be identified. The first event (1) whose s_{hmax} is NW-SE, is Valanginian in age and deforms the basal portion of the lacustrine section (sedimentation of the rift phase—Pendência Formation). The second event (2), of Late Aptian age and coaxial with the first event, obliquely reactivated the normal faults of the rift phase in an inverse regime and doubled the transitional section (Alagamar Formation). The third pulse (3), of a post-Campania age and s_{hmax} approximately N-S, regionally folded the post-rift section (Açu and Jandaíra Formations), reactivated normal faults of the rift phase and formed inverse NE-SW faults of lower angle involving the basement. Data on outcrops from the Barreiras Formation and the Açu Formation revealed the existence of the fourth event (4) post-Miocene tectonic inversion (Person Neto et al. 2008).

The last two of these events (third and fourth pulse) are responsible for deformations in the post-rift phase of the Potiguar Basin. These deformations are characterized by vertical dislocations that resulted in a variation that can exceed the 300 m of the Cretaceous–Neogene. These dislocations resulted in the formation of antiforms and synforms that embed the valleys and are expressed on the surface in the form of structural elevations (Serra do Mel and Serra de Mossoró) and structural depressions (valley of the Mossoró River and the valley of the Açu River) all in NE-SW orientation. This compression caused an antiform deformation in the post-rift section with maximum altitudes located more than 50 m above the maximum eustatic post-Cenomanian elevation (Haq et al. 1987). However, analysis of these elevations needs to consider the general flexure of the Atlantic Margin in the Cenozoic, because as the top of the post-rift section is above 200 m in the Serra de Mossoró, it is located below sea level on the coast.

The influence of these variations on the elevations of the post-rift section in the relief occurs in the formation of structural highs and topographic lows, both of NE-SW orientation. In the depressions between the antiforms, there are valleys of an

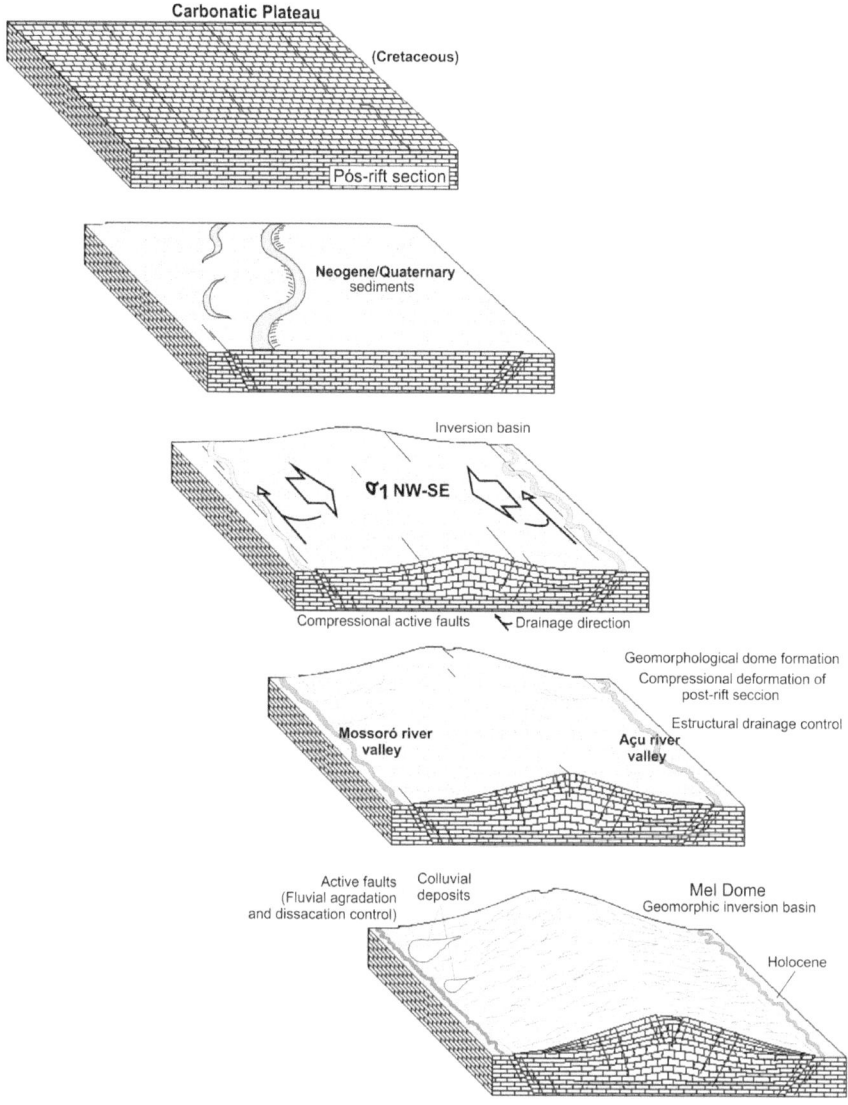

Fig. 5.6 Dome formation in the central part of Potiguar Basin by compressional tectonics in Ceno-zoic

NE-SW direction. These valleys correspond to topographic channels embedded in the drainage network and form the low river plains of the equatorial Atlantic margin. This structural conditioning exercises influence over the Quaternary sedimentation causing confined plains in an NE-SW direction. In altimetric terms, the representation of these structural lows is expressed in the form of depressions situated at an altitude of 0 m. In the valley of the Mossoró River, the elevation of 0 m penetrates up to about 30 km into the continent and in the Açu River valley as far as 8 km.

The patterns of structural deformation have a predominantly NE-SW orientation and these directions condition the main valleys forming channels parallel to the direction of the faults. These faults affect the Neogene and Quaternary coverings, as deformations in non-cohesive sediment and anomalies in the drainage. Examples of these anomalies are found in the main river channels, specifically in the stretches where they cross active fault zones. The structural control of the drainage occurs from the adaptation of the main channels to the NE-SW faults, while the NW-SE faults control the channels of the first and second order.

The collection of bibliographic data on Cenozoic reactivation in the Potiguar Basin, and the morphotectonic context of the compressional basins associated with geomorphological data obtained in the field and with remote sensing, allow the proposition that the uplift of the Potiguar Basin and the formation of domical structures in its central portion may have occurred in response to Miocene–Quaternary ($\sigma 1$) E-W and NW-SE compression.

This compression, resulting from the current $\sigma 1$ stress field with an NW-SE orientation of the Brazilian Atlantic equatorial margin reactivated formerly normal faults under a compressive regime, thus creating the sequence of elevations in an NE direction, demarcated by the Serra de Mossoró and the Serra do Mel, and the crests that demarcate linear interfluves. In particular, the Serra do Mel still preserves the sedimentary cover from the Quaternary sedimentation event which is an important yardstick for the minimum age of the uplift, in this case, the late Pleistocene. However, the Serra de Mossoró makes up a residual elevation resulting from an older uplift, supported by a top of silicified sandstone and therefore is resistant to differential erosion, which has maintained it as a topographic outcrop in the center of the basin.

These outcrops, either in a domical form or partially worn by erosion, have been interpreted as inversion features and are recognized by their well-known NE-SW orientation. This direction of the faulting is the main structural direction, affecting the Precambrian basement that in turn defines the direction of the Potiguar rift in the Cretaceous. Subsequently, the post-rift reactivation events reactivated the older systems of faults and deformed the Cretaceous and Cenozoic sedimentary sequences, including the Quaternary.

References

Ascione A, Romano P (1999) Vertical movements on the eastern margin of the Tyrrhenian extensional basin. New data from Mt. Bulgheria (Southern Apennines, Italy). Tectonophysics 315(1):337–356, 31

Bezerra FHR, Vita-Finzi C (2000) How active is a passive margin? Paleoseismicity in Northeastern Brasil. Geology 28:591–594

Bezerra FH, Amaro VE, Vita-Finzi C, Saadi A (2001) Pliocene-quaternary fault control of sedimentation and coastal plain morphology in NE Brazil. J South Am Earth Sci 14:61–75

Bezerra FHR, Neves BBB, Correa ACB, Barreto AMF, Suguio K (2008) Late Pleistocene tectonic-geomorphological development within a passive margin—the Cariatá trough, northeastern Brazil. Geomorphology 01:555–582

Bezerra FH, do Nascimento AF, Ferreira JM, Nogueira FC, Fuck RA, Neves BB, Sousa MOL (2011) Review of active faults in the Borborema Province, Intraplate South America Integration of seismological and paleoseismological data. Tectonophysics (Amsterdam) 510:269–290

Borges WRE (1993) Caracterização estrutural da porção sudoeste do Rift Potiguar, Brasil. 146 f. Dissertação (Mestrado em Geologia)—Programa de Pós-graduação em Geologia, Universidade Federal de Ouro Preto, Minas Gerais

Cipollari AP, Domenico Cosentino A, Elsa G (1999) Extension- and compression-related basins in central Italy during the MessinianLago-Mare event. Tectonophysics, 315:163–185

Cremonini OA, Kraner GD (1995) Reativação mesozóica da Bacia Potiguar. In: Simpósio De Geologia Do Nordeste, vol 6, Natal. Anais do VI Simpósio de Geologia do Nordeste, Natal, pp 181–184

de Castro DL, Bezerra FH, Sousa MO, Fuck RA (2012) Influence of Neoproterozoic tectonic fabric on the origin of the Potiguar Basin, northeastern Brazil and its links with West Africa based on gravity and magnetic data. J Geodynam 54:29–42

Doré AG, Lundin ER, Kusznir NJ, Pascal C (2008) Potential mechanisms for the genesis of Cenozoic domal structures on the NE Atlantic margin: pros, cons and some new ideas. In: The nature and origin of compression in passive margins. Geological Society, London, Special Publications

Haq BU, Hardenbol J, Vail PR (1987) Chronology of fluctuating sea levels since the Triassic. Science 235:1156–1167

Jackson J, Norris R, Youngson J (1996) The structural evolution of active fault and food system in central Otago, New Zealand: evidence revealed by drainage patterns. Struct Geol 18:217–234

Lugt IR, Wees JD, Wong TH (2003) The tectonic evolution of the southern Dutch North Sea during the Palaeogene: basin inversion in distinct pulses. Tectonophysics 373:141–159

Maia RP (2012) Geomorfologia e Neotectônica no Vale do Rio Apodi-Mossoró NE/Brasil. Tese de Doutorado. Programa de Pós-graduação em Geodinâmica e Geofísica. PPGG–UFRN. Natal, RN

Maia RP, Sousa MO, Bezerra FH, Neto PX, de Moura Lima EN, Silva CC, Santos RD (2012) A importância do controle tectônico para a formação do relevo cárstico na Bacia Potiguar - Nordeste do Brasil. Revista Brasileira de Geomorfologia

Matos RMD (1992) The northeast Brazilian rift system. Tectonics 11(4):766–791

Matos RMD (2000) Tectonic evolution of the equatorial South Atlantic. In: Mohriak WU, Talwani M (eds). Atlantic rifts and continental margins. AGU Geophys Monogr 115:331–354

Moura-Lima EN, Sousa MO, Bezerra FH, de Aquino MR, Vieira MM, Lima-Filho FP, da Fonseca VP, do Amaral RF (2010) Sedimentação e deformação Tectônica cenozóicas na porção central da Bacia Potiguar. Geologia USP. Série Científica, vol 10

Muñoz N, Charrier N, Jordan T (2002) Interactions between basement and cover during the evolution of the Salar de Atacama Basin, northern Chile. Revista Geológica do Chile, Santiago 29(1):55–80

Nogueira FC, Bezerra FHR, Fuck RA (2010) Quaternary fault kinematics and chronology in intraplate northeastern Brazil. J Geodyn 49:79–91

Pessoa Neto OC, Soares UM, Silva JGF, Roesner EH, Florencio CP, Souza CAV (2007) Bacia Potiguar. Boletim Geociências da Petrobras, Rio de Janeiro 15(2)

Quintana Q, Alonso JL, Pulgar J, Fernandez LRR (2006) Transpressional inversion in an extensional transfer zone (the Saltacaballos fault, northern Spain). J Struct Geol 28:2038–2048

Reis ÁF, Bezerra FH, Ferreira JM, Nascimento AF, Lima CC (2013) Stress magnitude and orientation in the Potiguar Basin, Brazil: Implications on faulting style and reactivation. J Geophys Res Solid Earth 1: n/a-n/a

Rossetti DF, Bezerra FH, Góes AM, Brito-Neves BB (2011) Sediment deformation in Miocene and post-Miocene strata, Northeastern Brazil: evidence for paleoseismicity in a passive margin. Sed Geol 235:172–187

Rossetti DF, Bezerra FH, Dominguez JM (2013) Late Oligocene-Miocene transgressions along the equatorial and eastern margins of Brazil. Accepted for publication: Earth Sci Rev

Sousa MOL, Bezerra FHR (2005) As tensões Tectônicas Campanianas-Cenozóicas na Bacia Potiguar, Brasil. In: Simpósio De Geologia Do Nordeste, vol 21, Recife. Anais do XXI Simpósio de Geologia do Nordeste, Recife: SBG, pp 329–330

Turner JP, Williams GA (2004) Sedimentary basin inversion and intra-plate shortening. Earth Sci Rev 65:277–304

Underhill JR, Paterson S (1998) Genesis of tectonic inversion structures: seismic evidence for the development of key structures along the Purbeck-Isle of wight disturbance. J Geol Soc London 155:975–992

Ventisette CD, Montanari D, Sani F, Bonini M (2006) Basin inversion and fault reactivation in laboratory experiments. J Struct Geol 28:2067–2083

Williams GD, Powell CM, Cooper MA (1989) Geometry and kinematics of inversion tectonics. Geol Soc London, Special Publications, vol 44, pp 3–15

Zanchi A, Berra F, Mattei M, Ghassemi MR, Sabouri J (2006) Inversion tectonics in central Alborz Iran. J Struct Geol 28:2023–2037

Chapter 6
Karstic Geomorphology

Abstract This chapter presents a deductive analysis of the geomorphological evolution of Northeastern Brazil. This analysis faces the need to update the interpretations of landform evolution, which should take into account the work carried out by rivers, to revisit the classical literature and contrast their interpretations with more recent morphotectonic research. It can be said that the dissection and deposition that occurred in the alluvial systems at the regional scale exhibit strong structural control. Fault reactivations are interpreted here as widely responsible for Neogene deformations in the study area. These reactivations also control the morphology and trigger processes of dissection and deposition. They are important mechanisms that should be considered in morphotectonic interpretations.

Keywords Geomorphology · Northeast · Neotectonics · Alluvial valleys

The term karstic is used to describe all the erosional features caused by dissolution, corrosion, and abrasion processes that occur with soluble rocks (De Waele et al. 2009; Klimchouck, 2009). Karstic formations are noteworthy for their runiform shape (Kohler 1995). These structures include walls etched and corroded by chemical weathering, caves, sinkholes, doline lakes, and pavements carved in carbonates (Carvalho Júnior et al. 2008).

In Northeastern Brazil, the development of karst morphologies is influenced by climatic factors. An average of 8 months of the year of drought and low levels of air humidity limit chemical morphogenic processes. In the current climate, this feature does not favor processes of carbonate dissolution related to chemical weathering. However, this limiting factor is not enough to make the region devoid of karst features since in some areas such as the Potiguar Basin, located in the states of Rio Grande do Norte and Ceará, a number of karst reliefs can be identified. In this case, the karst morphology and geological indicators of past climates, such as calcareous tufa deposits, are important paleoclimatic indicators (Auler et al. 2005; Boggiani et al. 2002). These indicators can provide important data on the evolution of karst landscapes currently subject to a semiarid climate. According to the database of natural underground cavities in the State of Rio Grande do Norte, managed by the CECAV nucleus (Study Centre, Protection and Cave Management), 563 cavities are recorded, namely 469 caves, 54 shelters, 36 abysses, and 4 sinkholes in Rio Grande

R. Maia and F. Bezerra, *Structural Geomorphology in Northeastern Brazil*,
SpringerBriefs in Latin American Studies,
https://doi.org/10.1007/978-3-030-13311-5_6

Fig. 6.1 Skinhole in Jandaíra Formation—Potiguar Basin

do Norte. This high speleogenic potential puts Rio Grande do Norte in seventh place
in Brazil and second in the northeast with the most known cavities, just behind
Bahia. Of these cavities, 91.5% are located in the limestone Jandaíra Formation in
the Potiguar Basin (Cruz et al. 2010). This scenario is due to the extensive carbonate
platform of the Jandaíra Formation, which occupies a large area of the basin and is the
largest exhibition of limestone rocks in Brazil. This carbonate surface expresses itself
through a set of deformations, the brittle events of the post-rift age. In the Potiguar
Basin, the Cretaceous stress fields were responsible for the emergence of different
fault systems in a predominantly NW-SE and NE-SW direction. These faults occur
on the surface and affect the top of the post-rift section of the basin, thus exerting an
important influence on its geomorphological evolution.

This morphology occurs mainly in the form of caverns, sinkholes, karstic valleys, canyons, and pavements. These are a set of reliefs developed according to the
direction of the main systems of regional faults that affect the Potiguar Basin. Currently, these systems of faults have an important influence on the geomorphological
evolution of this basin, as it controls the drainage and directs the dissection and
deposition of Quaternary sediments (Moura-Lima et al. 2011a, b; Maia and Bezerra
2012) (Figs. 6.1 and 6.2).

Thus, the purpose of this chapter is to characterize the systems of outcrop faults in
Cretaceous carbonates of the Potiguar Basin and show the importance of fault systems

Fig. 6.2 Small dry chanell in Jandaíra Formation—Potiguar Basin

in the conditioning of karst features, as well as their implications for the origin of caves. The study area is the onshore portion of the Potiguar Basin, located in Brazil's far northeast, in Rio Grande do Norte and the east of Ceará, between 35° and 38° west longitude and between the 05° and 06 latitude south. The area to be analyzed in this chapter is defined by the occurrence of the Jandaíra Formation carbonates, the post-rift sequence of the Potiguar Basin. This lithostratigraphic unit is up to 700 m thick and is extensively distributed in the basin. The outcrops usually range from 1.0 to 6.0 m thick and are observed mainly in banks and beds of dry drainage, pavements, and mines for mineral exploration. Specifically, they correspond to the western portion of the basin, the location of the largest outcrops of the Jandaíra Formation and therefore several features of karst relief.

6.1 Geological and Environmental Characterization

The Potiguar Basin is genetically related to a series of inland basins from the Lower Cretaceous age, which compose the Rift System of northeastern Brazil (Matos 2000a, b; de Castro et al. 2012). The basin's structural framework consists of a set of asymmetric grabens of an NE-SW direction separated by horst from the basement. This structural construction called the Potiguar Rift stemmed from the NE-SW trend of the ductile structures of the crystalline basement during the Lower Cretaceous (Nobrega et al. 2005; Castro et al. 2012). This rift is limited to the east and west by the Carnaubais and Areia Branca faults, respectively, which constitute a dual system of normal faults, which would have developed during the Mesozoic reactivation of the Neoproterozoic shear zones, from the opening of the Atlantic Ocean (Matos 1992).

In the Potiguar Basin from the Cenozoic, the tectonic features have a smaller regional expression, not determining the emergence of rifts and sedimentary basins, as occurred during the Mesozoic. However, in the Cenozoic, there were events such as the reactivation of important faults (Carnaubais and Afonso Bezerra Fault Systems) (Moura-Lima et al. 2010; 2011a, b), folding with large wavelength and axes oriented in an N-S direction, resulting from EW compressive stresses affecting this basin in the Paleogene (Cremonini 1993), and tectonic reactivation associated with basic intrusions related to the Macau Formation (Knesel et al. 2011).

The faults of the central and west portion of the Potiguar Basin are already well known. They mainly included the fault systems of Afonso Bezerra (NW direction) and Carnauba (NE direction). The Alfonso Bezerra system consists of transfer faults, while the Carnaubais fault system results from the reactivation of the Portalegre shear zone that occurred in the Atlantic opening (Castro et al. 2012). These faults play an important role in defining the drainage patterns that constitute one of the key parameters in identifying neotectonic movements. These movements are mainly induced by faults reactivated during the Cenozoic, from the compressive regime located on the edge of the Brazilian equatorial Northeast (Bezerra and Vita Finzi 2000).

Two stress fields that operated in the Paleogene and another from the Neogene to the Quaternary were identified by Bezerra and Vita Finzi (2000) and Bezerra et al. (2011), from the analysis of field data, fault mechanisms, and breakout. The first field was characterized by compressive stresses of an approximately N-S direction and E-W distension; the second field is characterized by the direction of compression ranging from NW-SE and NE-SW distension; in the central part of the Potiguar Basin, an E-W compression and N-S distension in the eastern portion of the basin. This last field affects all the lithostratigraphic units and is the only field of stresses whose effects are observed in the Barreiras Formation and Quaternary sediments. According to these authors, the kinematics of the faults are characterized by NE-SW dextral and NW-SE sinistral strike-slip faults, normal faults, and E-W joints.

The drift phase of the Potiguar Basin occurred from the peak of a breach between the Cenomanian (99.6–93.6 Ma) and Turonian (93.6–88.6 Ma), marked by the drowning of the river systems of the Açu Formation as well as the implementation of an extensive carbonate platform called the Jandaíra Formation (Soares et al. 2003). The carbonates of the Jandaíra Formation are, in general, grayish (dolomitized limestone and dolomites) to pale cream (limestone) in color, with textures ranging from fine to coarse. Deposition occurred in the period ranging from Turonian (93.6–88.6 Ma) to Campanian (83.5–70.6 Ma) at the end of the formation period of the transgressive sequence in conditions of continental drift and open seas (Araripe and Feijó 1994).

The Jandaíra Formation is the most extensive outcropping of Phanerozoic carbonates in Brazil. The rocks of the Jandaíra Formation are a carbonate ramp that is found on virtually all the onshore portion of the Potiguar Basin. This carbonate ramp was submitted, during and after deposition, to various episodes of uplift and erosion causing subaerial exposure that resulted in intense karstification and dissolution. The karstification of the Jandaíra Formation has important structural conditioning, especially the tectonic stresses that acted after its deposition, at the end of the Campanian. The post-Campanian tectonic is imprinted on the karst of the Jandaíra Formation as faults, fractures, and raised blocks promoting the rejuvenation of the hydrodynamic profile, favoring and conditioning the speleogenetic in several stratigraphic levels. The karstification developed in the Jandaíra Formation is essentially epigenetic and the planes of faults and fractures serve as conduits for the percolation of rainwater that promotes the dissolution the carbonate rock (Xavier Neto et al. 2008).

The analysis area is situated in the domain of BSW'h type semiarid climates, according to the Köppen classification (Radambrasil 1981); it is characterized by the predominance of high temperatures associated with a system of sporadic rains, mostly concentrated in the first four months of the year. According to Nimer (1989), the semiarid climate is influenced by the intertropical convergence zone, with the dry season from June to January and the rainy season from February to May. The rainfall system is mainly controlled by various mechanisms, notably cold fronts, the position of the Intertropical Convergence Zone (ITCZ) and the Upper Cyclonic Vortices, and easterly waves (Noble 1994). The average annual rainfall is around 700 mm and annual temperatures are around 27 °C, with a minimum of 21 °C and a maximum of 36 °C. The relative humidity throughout the year in the region is on an average of

70% and follows the precipitation curve, with higher values observed from February to May and lower values from June to January (IDEMA 2012).

From the hydrographic point of view, the area is located in the lower course of the basin of the Apodi-Mossoró River (Fig. 4.12). In this sector, the decrease in density of the channels is due to the increased permeability of the substrate composed of the carbonates of the Jandaíra Formation. Differently to what occurs in the upper reaches, to the south high drainage density results from its relationship with the Precambrian basement.

In the lower course, the NW-SE (Afonso Bezerra Fault System) and the NE-SW (Carnaubais Fault System) fault systems of the Potiguar Basin influence a parallel drainage in an NE-SW orientation for main channels (third and fourth order) and NW-SE for tributaries (first and second order). The effects of brittle structures on the drainage are important evidence of structural control and post-rift reactivation.

6.2   Post-rift Tectonic and Faults in the Potiguar Basin

The Cretaceous and Cenozoic tectonic that acted in the study area is represented by macro and mesoscopic structures identified in the rocks of the Jandaíra Formation. The layout of the structures in the research area behaves like swarms of lineaments from meters to a few kilometers with significant speleogenic potential. The analysis of products using remote sensing shows that the lineaments drawn from digital images are oriented in NE-SW, E-W, and N-S directions. However, there is a predominance of the NW-SE direction, related to the Afonso Bezerra, Poço Verde-Caraúbas, and Algodão-Juazeiro fault systems.

The extraction and interpretation of these lineaments, associated with the data obtained in the field, allowed the verification of the expression on the surface of these fault systems. The major lineaments of NW-SE orientation have clear control in the occurrence of Quaternary deposits. Ridges and drainage valleys, which expose rocks of the Jandaíra Formation, are also aligned according to this orientation. The occurrence of silicification in the limestone of the Jandaíra Formation produces lineaments marked by topographic highs. The NW-SE lineaments are mainly marked by secondary drainage channels of the Apodi-Mossoró and Açu Rivers.

The NE-SW lineaments are related to the Carnaubais Fault system and other parallel faults. These faults are strongly indicated in the control of the region's two great rivers (Apodi-Mossoró and Açu Rivers), which expose rocks of the Jandaíra Formation. The parallel drainage oriented in an NE-SW direction for main channels (third and fourth order) and NW-SE for tributaries (first and second order) is also conditioned by NE-SW and NW-SE. Several drainage elbows interrupt the NE-SW parallelism, forming NW-SE segments and are an important evidence of Quaternary reactivation (Maia and Bezerra 2012). These faults document a field of stresses related to NE-SW compression and NW-SE extension compatible with a sinistral strike-slip regime for the faults of the Afonso Bezerra, Poço Verde-Caraúbas, and Algodão-Juazeiro systems. It is worth noting that some structures not observed on satellite images are present on the scale of outcrop or detail photos and from the "karst" perspective are equally important because they connect different sets of fractures, both horizontally and vertically. Based on field observations, it is observed that all the outcropping limestone appears deformed in a brittle regime, marked by breaks ranging from centimeters to hundreds of meters.

Due to major faults, the limestone was affected by silicification processes and underwent a further process, hydraulic brecciation (Moura Lima et al. 2011a, b). The silicification and brecciation of carbonate rocks are easily recognized in the field, as these processes give the rock more resistance to erosion, which makes these features stand out amid a relatively flat relief, characteristic of the interior of sedimentary plains.

Two types of sedimentary deposits occur along dissolved faults. The first is represented by the gaps formed by collapse (collapse or chaotic breccias) and consists of fragments falling from the wall of the faults or the ceiling of caves after karstification. The collapse breccias are poorly selected and are usually supported clasts. The clasts vary from boulders to pebbles and the matrix is composed of granulometric fragments of sand and silt. The second type of fill are detrital alluvial sediments derived from nearby rivers and the degradation of escarpment faults, which were transported in traction or suspension to be deposited along faults opened by dissolution.

6.3 Karstic Morphology

The onshore portion of the Potiguar Basin is a cuesta stretching from the west segment of Rio Grande do Norte and the eastern end of the State of Ceará, with the front facing S and SW and the reverse side facing NE at an angle of $0.1°$. In the front, the steep cornice is supported by carbonate rocks from the Jandaíra Formation with somital of approximately 140 m. The solubility of carbonate rocks in this area does not favor the formation of talus deposits at the interface between the front and the surrounding peripheral depression. At the base, the front forms a ramp with very variable inclination shaped in the Açu Formation sandstones. This sector presents incipient dissection performed by channels of the first and second order of anaclinal and ortoclinal types toward the *Sertanejo* Depression limiting the area of the Precambrian basement and the sedimentary basin (Fig. 6.3).

The karstic relief may be defined by the occurrence of external zones or exokarstes, zones of contact of the rock with soil, or epikarstes, and the subterranean zones called endokarstes (Klimchouck 2009). Of these, each can be further divided into positive and negative forms. Positive karst morphologies are all those resulting from the lateral erosion and corrosion remaining as trace, a testimony of a former superficial eroded karst.

In the Potiguar Basin, the expression on the surface of the brittle deformation related to recent stress fields that influenced the evolution of the karst relief in all forms, are of the endokarstic or exokarstic types. This influence happens through the structural control of drainage, favoring those sectors of the localized development

Fig. 6.3 Apodi plateau—Potiguar Basin

Fig. 6.4 Extended Fracture examples (Soledad Lajedo, Apodi—RN) (*Source* Maia et al. 2012)

of ruinform morphologies. In this respect, the structural and tectonic controls are important in the morphological and directional conditioning of various karst features such as sinkholes, caves, and valleys (Auler et al. 2005).

A result of the lithostructural conditioning has been the formation of incised valleys oriented according to the directions of faulting (Fig. 6.4).

The confluence of these channels originates incised valleys where endokarstic-type forms such as caves can be identified and also exokarst ones like sinkholes and canyons. This is a set of reliefs developed according to the directions of the major regional fault systems that currently exert structural control on the drainage and hence, the dissection.

The morphological pattern of the channels formed from the fluvial dissection is linear along the faults, with the larger ducts occurring at their intersection.

Under the pavements of the Jandaíra Formation, drainage patterns are mostly of the parallel and trellis type. These patterns can be identified by means of remote sensing in the form of negative topographic lineaments expressed in the field in the form of incised valleys and lapiás. However, their terminations are difficult to identify since the runoffs commonly drain into sinks; an endogenous circulation pattern.

In the area of pavements, the carbonates of the Jandaíra Formation form extensive Lapiás fields individualized by canyons (extended fractures). In this sector, the carbonates occur mainly in the form of limestone, dolomitized limestone, and dolomite (Bezerra et al. 2012) with texture ranging from coarse to fine. This textural pattern is expressed on the walls of the extended fractures as a set of erosional features that follow the stratification planes. These features are more pronounced in thinner and therefore more friable facies. Linear horizontal cavities originate from these facies that start to show geomorphologically both the textural classes and the stratigraphic architecture of carbonate outcrops. Whereas on the surface, the directions of the Lapiás take on two distinct patterns. The first is a diffuse type, with no apparent control of the morphology by the lithology or structure. The second pattern is characterized by reminiscent linear crests that adapt to the lines represented by the structural weakness fractures and sets of joins. In this case, the lapiás are characterized by a sequence of parallel ridges individualized by recessed linear features that channel the runoff, accelerating the chemical corrosion in these sectors.

The maintenance of the crests occurs due to the lithological difference since they are fractures filled with calcite and this is a resistance factor to chemical corrosion. These crests occur mostly on a centimeter scale reaching a decimeter scale (Fig. 6.5).

The Jandaíra Formation is marked by a high degree of tectonic deformation. The origin of the caves is directly linked to the faults of the Potiguar Basin evidenced by the direction of the main galleries. As an example, there is the Arapuá Cave which coincides with the regional trend, or is distributed according to an NW-SE and NE-SW direction. Coupled with the structural factor, the caves are formed because limestone is a fragile soluble rock susceptible to acidity and when in an extremely fractured and porous environment, such as the Jandaíra Formation, the new surfaces created from fracturing become potential karstification surfaces (Gomes et al. 2011). In such cases, the formation of channels is controlled by the solubility of the rock and its structural pattern. Based on field evidence, it is clear that the presence of NW-SE and NE-SW structures exerts strong control and influence directly on the formation of underground cavities, since these directions coincide with the preferred direction of the caves.

The orientation of faults, fractures, and joints also influence the work of dissection and the formation of exokarstic features. These structures facilitate the access of water inside the rocks accelerating their chemical wear. Thus, fractured zones potentiate the erosive action forming karst features oriented as per the directions of the faulting. In the Potiguar Basin, the NE-SW and NW-SE direction of fault systems document the effects of the last Cenozoic stress fields.

Fig. 6.5 Overview of pavement in displaying extensive surface Lapiás field. On the slopes of extensive fractures, erosional depressions resulting from differential erosion can be identified

The effects of post-rift tectonic reactivations can be observed from geological indicators such as faults filled with tufa limestone or calcite and vertical stylolites. From geomorphological point of view, the normal faults, collapse or dissolution sinkholes and incised valleys oriented in fault lines are the main tectonic morpho indicators of the basin.

Over the carbonates of the Jandaíra Formation, the faults condition the development of an angular drainage system of the lattice or rectangular type. This drainage dissects the carbonates causing incised valleys of NW-SE and NE-SW orientation. The metric fractures, in turn, embed the current hydrographic network (Fig. 6.6).

The points that exhibit evidence of karstification are devoid of soil cover. This fact reveals that the erosion processes associated with these areas supplanted the pedogenesis that occurs extensively on the surface of the Potiguar Basin. In this regard, shallow Eutrophic Cambisols and Rendzinas soils are distributed in the research area in places less affected by Cenozoic tectonics. As the fragile structures conditioned the erosive action preferentially according to the planes of faults and fractures, areas with more evidence of karstification had, in genetic terms, their pedogenesis overcome by physical and chemical morphogenesis represented by superficial and underground runoffs confined by the shallow brittle structures. Thus, the surface of the pavement has an extensive set of morphologies associated with fluvial dissection. Noteworthy

Fig. 6.6 (Photo A—Incised structural valley) (Photo B—Fracture in early stage of enlargement) (Photo C—Joins filled with calcite) (*Source* Maia et al. 2012)

Fig. 6.7 Structural karst features with orthogonal fracture systems

in this group is the small ravines carved by runoff, which give a rugose surface to the pavement.

Structural valleys are the most common of the numerous karstic features in the Potiguar Basin. These valleys result in lineaments that are reflected in the topography of the region as rectilinear features related to the regional fault systems reactivated in the Cenozoic. In this context, the post-rift deformation events were of great importance in the origin and evolution of karstic relief in the basin, since the pattern of structural deformation conditions the weathering action, mainly represented by fluvial dissection, resulting in linear morphologies reflecting in their direction the fields of Cenozoic tectonic stresses responsible for reactivation.

The current erosive processes are controlled mainly by physical morphogenesis conditioned by the semiarid climate as the chemical morphogenesis only occurs in the rainy season, limited to the period from February to May. During this period, a dense network of channels develops on the carbonates, adapting to the preexisting brittle structures as faults and sets of joints.

Fractures are the starting point for the development of secondary porosity (the fissure type) in this rock. It is through them that the water starts its slow and continuous breakdown of rock, leading in the long term to microporosity of the underground cavities (Figs. 6.7 and 6.8).

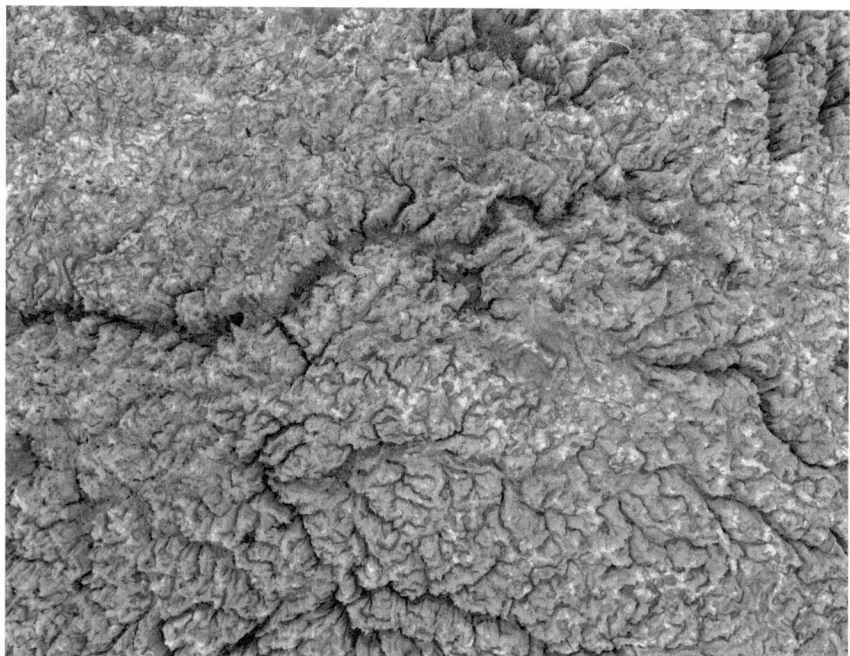

Fig. 6.8 Surface karst features without structural control

In addition to the structural control represented by the faults, fractures, and sets of joints, the epigenic control related to the baseline level means tectonic conditioning becomes even more apparent as the dissolution of carbonates only occurs with intensity in wet periods. In these periods, structural cracks in the carbonates channel the runoff that goes on to develop these incisions and from the geomorphological point of view, enhance the current structural framework.

This climatic context, where rainfall is concentrated in a short period of time, limits the dissolution of carbonates in the aqueous solution existing between carbon dioxide (CO_2), carbonic acid (H_2CO_3), bicarbonate ions (HCO_3^-), and carbonate (CO_3^2-) (Carvalho Júnior et al. 2008). Thus, the chemical morphogenesis associated with fluvial streams is limited to intermittent seasonal channels that have a high erosive competence and solid discharge in the rainy season and are dry valleys for the rest of the year.

In the Potiguar Basin, karst valleys can be found in different phases as shown in Figs. 4.15 and 4.18. These valleys were formed by the dissolution of carbonates in fractures that resulted in the linear lowering of the surface thus constituting incised valleys. These incised valleys are important indicators of post-rift reactivation since their direction follows the orientation of faults affecting the Jandaíra Formation in the Potiguar Basin. These failures, of an NE-SW and NW-SE direction have played an important role in developing karstic morphologies as the seasonality of drainage

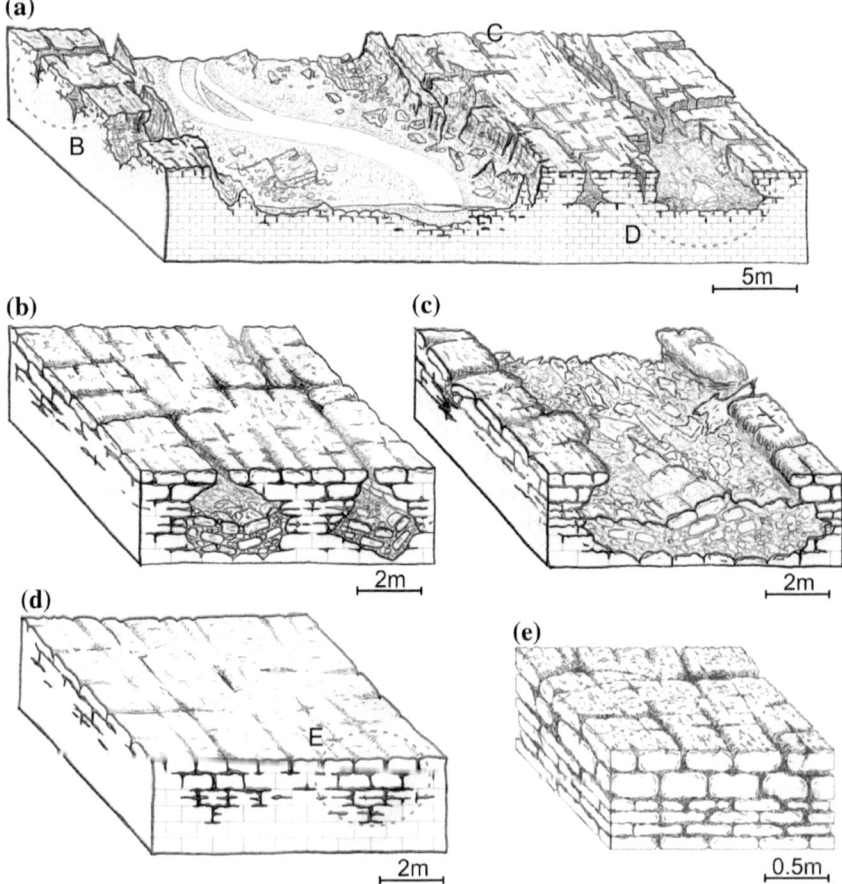

Fig. 6.9 Evolution of karst modeled from the fractured surface (*Source* Orildo Lima et al. 2017)

conditions the linear dissection of carbonates that start to exhibit a complex network of ravines and valleys incised into the surface (Fig. 6.9).

The evolutionary history of the Potiguar Basin is marked by tectonic reactivations which led to a complex fault system in the Jandaíra Formation. In many cases, these fault systems are geomorphological and are amenable to observation in the field, as well as satellite images and aerial photographs. Currently, such faults are also an important influence on the geomorphological evolution of the basin, which has come to show the current complex structural framework in the relief.

The post-rift tectonics have various evidence of the Cenozoic reactivation, expressed on the surface as relief and drainage features and in shallow structures identified in the field. These structures are invariably formed by the solubility of the carbonates in a structural pattern determined by the regional fault systems of the

Potiguar Basin. These fault systems, reactivated in the Cenozoic, have played an important role in defining the shape of the karstic basin.

This shape includes ample and extensive fractured pavements, Lapiás fields related to fractured pavements at older erosive stages, collapse sinkholes, incised valleys, and ravines developed from the enlargement of the fault planes.

References

Araripe PT, Feijó FJ (1994) Bacia Potiguar. Boletim de Geociências, RJ 8(1)

Auler AS, Piló LB, Saadi A (2005) Ambientes cársticos. In: Souza CRG, Suguio K, Oliveira AMS, Oliveira PE (Org.). Quaternário do Brasil. Ribeirão Preto: Holos Editora, v., pp 321–342

Bezerra FHR, Vita-Finzi C (2000) How active is a passive margin? Paleoseism Northeastern Brasil. Geology 28:591–594

Bezerra FH, do Nascimento AF, Ferreira JM, Nogueira FC, Fuck RA, Neves BB, Sousa MOL (2011) Review of active faults in the Borborema Province, Intraplate South America Integration of seismological and paleoseismological data. Tectonophysics (Amsterdam) 510:269–290

Bezerra FHR, Srivastava N, Souza MOL (2012) Relatório do Mapeamento Geológico Regional: Folha SB-24-X-D-I (1:100.000) Contrato CPRM-UFRN No 59/PR/08. Natal, RN

Boggiani PC, Coimbra AM, Gesicki AL, Sial AN, Ferreira VP, Ribeiro FB, Flexor JM (2002) Tufas Calcárias na Serra da Bodoquema, MS. In: Scobbenhaus C, Campos DA, Queiroz ET, Berbert-Born MLC (eds) Sítios Geológicos e Peleobiológicos (SIGEP) V01 Congresso Brasileiro de Geologia 31, SBG, Anais, pp 607–617

Carvalho Junior OA, Berbet-Born M, Martins ED, Guimarães RF, Gomes RAT (2008) Ambientes Cársticos. In: Florenzano TG (Org.). Geomorfologia: Conceitos e Tecnologias Atuais. 1ª ed. São Paulo: Oficina de Textos

Cremonini OA (1993) Caracterização estrutural e evolução da área de Ubarana, porção submersa da Bacia Potiguar, Brasil. Dissertação (Mestrado)—Universidade Federal de Ouro Preto, UFOP, Ouro Preto

Cruz JB, Bento MD, Bezerra FHR, Freitas JI, Campos UP, Santos DJ (2010) Diagnóstico Espele-ológico do Rio Grande do Norte. Revista Brasileira de Espeleologia, vol 1 N 1

de Castro DL, Bezerra FHR, Sousa, Maria OL, Fuck RA (2012) Influence of Neoproterozoic tectonic fabric on the origin of the Potiguar Basin, northeastern Brazil and its links with West Africa based on gravity and magnetic data. J Geodyn 54:29–42

de Waele JD, Plan L, Audra P (2009) Recent developments in surface and subsurface karst geo-morphology: an introduction. Geomorphology 106:1–8

Gomes IP, Veríssimo CUV, Bezerra FHR (2011) As fraturas e sua importância no controle da paisagem cárstica em calcários da Formação Jandaíra (cretácio da bacia potiguar), Felipe Guer-ra—RN. 310 Congresso Brasileiro de Espeleologia Ponta Grossa-PR, Anais, 21–24 de julho de 2011—Sociedade Brasileira de Espeleologia

IDEMA. Caracterização do clima, solo, vegetação, relevo, recursos hídricos e aspectos sócio-econômicos do município de Mossoró. IDEMA, Instituto de Desenvolvimento e Meio Ambiente do Rio Grande do Norte (2002). Disponível em: www.idema.rn.gov.br/perfil_g.asp. Acessado em fevereiro

Klimchouck A (2009) Morphogenesis of hypogenic caves. Geomorphology 106(1–2):100–117

Knesel KM, Souza ZS, Vasconcelos PMP, Cohen, Silveira FV (2011) Young volcanism in the Borborema Province, NE Brazil, shows no evidence for a trace of the Fernando de Noronha plume on the continent. Earth Planet Sci Lett 302:38–50

Kohler HG (1995) Geomorfologia Cárstica. In: Guerra AJT, Cunha SB (Orgs). Geomorfologia: Uma atualização de Bases e Conceitos. 2 ed. Rio de Janeiro. Bertrand Brasil

Maia RP, Sousa MO, Bezerra FH, Neto PX, de Moura Lima EN, Silva CC, Santos RD (2013) A importância do controle tectônico para a formação do relevo cárstico na Bacia Potiguar—Nordeste do Brasil. Revista Brasileira de Geomorfologia

Matos RMD (1992) The Northeast Brazilian rift system. Tectonics 11(4):766–791

Matos RMD (2000a) Tectonic evolution of the equatorial south Atlantic. Geophys Monogr AGU 115:331–354

Matos RMD (2000) Tectonic evolution of the equatorial South Atlantic. In: Mohriak WU, Talwani M (eds.) Atlantic rifts and continental margins. AGU Geophys Monogr 115:331–354

Moura-Lima EN, Sousa MO, Bezerra FH, de Aquino MR, Vieira MM, Lima-Filho FP, da Fonseca VP, do Amaral RF (2010) Sedimentação e deformação Tectônica cenozóicas na porção central da Bacia Potiguar. Geologia USP. Série Científica, vol 10

Moura-Lima EN, Sousa MO, Bezerra FH, de Castro DL, Damascena RV, Vieira MM, Legrand JM (2011) Reativação Cenozóica do Sistema de Falhas de Afonso Bezerra, Bacia Potiguar. Geociências (São Paulo), vol 30, pp 77–93

Moura-Lima EN, Bezerra FHR, Lima-Filho FP, de Castro David L, Sousa Maria OL, Fonseca VP, Aquino MR (2011b) 3-D geometry and luminescence chronology of Quaternary soft-sediment deformation structures in gravels, northeastern Brazil. Sed Geol 235:160–171

Nimer E (1989) Climatologia do Brasil. 421 p, IBGE, Rio de Janeiro, RJ, Brasil

Nobre P (1994) Clima e Mudanças Climáticas no Nordeste, Projeto Áridas, Ministério da Integração Nacional, V1, N1

Nóbrega MA, Sá JM, Bezerra FH, Neto JH, Iunes PJ, Guedes S, Saenz CT, Hackspacher PC, Lima-Filho FP (2005) The use of apatite fission track thermochronology to constrain fault movements and sedimentary basin evolution in northeastern Brazil. Radiat Measurem Amsterdam, 39:627–633

Radambrasil (1981) Folhas SB-24/25 Jaguaribe/Natal: Geologia e mapeamento geológico, Geomorfologia, Pedologia, Vegetação e Uso potencial da terra. Rio de Janeiro, 774 p. Levantamento de recursos naturais, 23

Soares UM, Rossetti EL, Cassab RCT (2003) Bacia Potiguar. Fundação Paleontológica Phoenix. Bacias Sedimentares Brasileiras, Ano 5, n 55

Xavier Neto P, Bezerra FHR, Silva CCN, Da E Cruz JBO (2000) condicionamento estrutural do carste Jandaíra e da espeleogênese associada pela tectônica pós-campaniana da Bacia Potiguar. In: Congresso Brasileiro de Geologia, 44, Curitiba, Anais SBG, CD-ROM

Chapter 7
Neotectonics and River Valleys

Abstract This chapter presents a deductive analysis of the geomorphological evolution of northeastern Brazil. This analysis faces the need to update the interpretations of landform evolution, which should take into account the work carry out by rivers, to revisit the classical literature and contrast their interpretations with more recent morphotectonic research. It can be said that the dissection and deposition that occurred in the alluvial systems at the regional scale exhibit strong structural control. Fault reactivations are interpreted here as widely responsible for Neogene deformations in the study area. These reactivations also control the morphology and trigger processes of dissection and deposition. They are important mechanisms that should be considered in morphotectonic interpretations.

Keywords Geomorphology · Northeast · Neotectonics · Alluvial valleys

Neotectonic refers to the study of the deformation caused by the latest stress fields. In the geomorphological approach, neotectonic has proven to be a powerful mechanism of morphogenetic and morpho-evolutionary analysis. Although the relation between tectonics and relief is common in Geomorphological studies, these relationships are extremely incipient or absent when the time scale analyzed is the present, especially the Quaternary.

According to Schumm et al. (2000), despite the practical significance of tectonic effects on fluvial environments, only a small number of studies considered such effects before the 1980s. Variations in the morphology of fluvial channels were interpreted, for example, as a consequence of variations in solid discharge and load type, thus hindering the detection of the effects of tectonic activity.

In Brazil, from the 1970s, several researchers linked to the Geotectonic and especially to morphotectonic, began to turn their interests to the tectonic activities that have occurred since the end of the Neogene to the Quaternary, evidenced by the morphology of the current relief and geological structures. Another factor that began to draw the attention of Geosciences in Brazil was the earthquakes that occurred more frequently in the northeast in the 1980s. Such phenomena have been reported since the last century, but the idea that Brazil was tectonically stable has caused the scientific community, in general, not to relate earthquakes these to global tectonics.

© The Author(s), under exclusive license to Springer Nature Switzerland AG 2020   101
R. Maia and F. Bezerra, *Structural Geomorphology in Northeastern Brazil*,
SpringerBriefs in Latin American Studies,
https://doi.org/10.1007/978-3-030-13311-5_7

The growing interest in the topic has changed this way of thinking. So, for those who are studying the geological–geomorphological processes occurring from the Neogene, it is clear that the current tectonics are one of the main mechanisms controlling these processes, as well as the morphology of the relief they have shaped (Lima 2000).

Currently, there are numerous geomorphological studies that attribute an increasing part of the explanations of the forms and morphogenesis to the tectonic factor (Saadi 1998). Thus, it has become increasingly clear that in addition to the paleoclimatic framework and the configuration of the basement, the current tectonic is of great importance in defining the evolutionary models, especially fluvial.

In the Brazilian Northeast, evidence of tectonic Post-Pliocene activity was observed by Bezerra and Vita-Finzi (2000), from its expression in several indicators, notably the structural control of drainage, liquefaction structures in fluvial sediments, and deformations and faults in Neogene rocks.

The configuration of the relief organized around the Borborema Plateau (Fig. 5.5) makes this an important disperser of drainage in the Northeast (Ab know 1969), where a dense drainage network is responsible for intense dissection. The rivers form valleys embedded with a preferred EW and NE-SW direction and exhibit generally rectilinear courses interspersed with meandering sectors, showing an adaptation to the preexisting geological structures such as faults and shear zones (Bezerra et al. 2001).

Sedimentary environments (Cretaceous Basins and sedimentary units) have revealed important indicators on how the Cenozoic processes of reactivation of faults have conditioned the evolution of the relief associated with these areas. In sedimentary environments, the repercussions of these fault reactivations can be observed from the direction of dissection and the resulting deposition of Quaternary sediments. Thus, the geomorphology of the northern Northeast, both in the Precambrian basement and the Paleozoic–Mesozoic Basins and Cenozoic deposits is conditioned by morphostructural structure. In this respect, the major river basins of the northern Northeast are directly conditioned by faults and ductile shear zones. Examples of this drainage include the Acaraú River in the north of the state of Ceará, the valley of which is embedded over the Transbrasiliano lineament; the Jaguaribe River develops its lower and middle course over a homonymous fault in a depression between the Senador Pompeu and Portalegre shear zones; the valleys of the Apodi-Mossoró and Piranhas-Açu Rivers, both embedded in the system of nonoutcropping faults of the rift phase in the Potiguar Basin. In the last case, the reactivation of the faults of the rift phase is responsible for the formation of the topographical highs and lows of the Potiguar Basin, thus influencing the configuration of the drainage.

Residual crests aligned along the main tectonic directions, edges of sedimentary plains affected by uplift and crystallines modeled by differential erosion, residual massifs individualized by flattened depressions where erosive processes have supplanted gradational ones and coastal plains shaped by eustasia, make up the complex mosaic of the northeastern landscape, documenting important episodes of its morphotectonic and paleoclimatic evolution. Among these units, the coastal plains, river valleys, and pre-coastal levels reveal important indicators of neotectonic events from various indicators.

In this sense, the present chapter proposes an analysis of the geomorphological significance of the neotectonic of the Northeast of Brazil, with an emphasis on the large fluvial systems. From the morphotectonic interpretation, there will be a focus on their evolutionary conditionings and their possible correlations with current seismic activity.

7.1 The Geomorphology of the Northeast: Genetic Aspects

Models of the geomorphological evolution of the Brazilian Northeast have been elaborated by different authors in the second half of the twentieth century. From an analysis of the hydrographic network, climatic variations and the weathering profiles situation on different elevations, Dresch (1957) identified three leveling paleosurfaces. Demangeot (1960) identified four paleosurfaces attributing an erosive event succeeding every epeirogenic phase. Based on the study of geological/geomorphological profiles, Ab Sáber (1960, 1969) suggested the existence of five paleosurfaces in the Northeast. Such surfaces would be the result of a complex interaction between climatic changes and tectonic processes where pedogenetic phases of hot and humid weather alternate with morphogenetic stages of hot and dry weather, with violent and sporadic rains, where pediplanation processes are in force. The application of this theory enabled the development of a Geomorphology of the Quaternary, with scientists from different fields of knowledge addressing the issue without, however, there being a precise definition of the methodological approach.

To the northeast, the model based on the occurrence of post-Cretaceous epeirogenies accompanied by phases of dissection and pediplanation conducted by dry climates was widely broadcast by Ab Sáber (1960), Bigarella (1994, 2003), Andrade and Lins (1965) Mabesoone and Castro (1975), among others.

These authors acknowledge the existence of terraced surfaces, resulting from planation stages resulting from erosion processes, due to the uplift of a continental core. Thus, the sedimentary sequences of the Mesozoic and Cenozoic would be the result of erosion resulting from uplift and consequently, of the lowering of the regional base.

The model proposed by King (1956) is based on the fact that the relief has a cyclical character but is not analogous to Davis' (1899) erosion cycle, as the pediplanation processes are often interrupted by phases of uplift. Applying this model to the northeast is based on the idea of planation and the development of young surfaces as a result of continental flexure in the Northeast of Brazil.

However, according to Saadi and Torquato (1994), the morphostructural evolution of Northeastern Brazil is based on the occurrence of major crustal cambering. These authors believe that the differences are between those who propose a swelling in the range of the Northeast core and those who propose the occurrence of various points or axes of crustal elevation, spatially related to major regional tectonic directions.

In this sense, the linear erosion processes sectioning the river valleys would be triggered by an uplift of polygenic origin. Such a process would cause slopes which, when subjected to drought would retreat laterally maintaining their altimetry interpreted as paleosurfaces. The role of the tectonic is demonstrated in promoting the variations in the baseline levels, inducing dissection.

With the advent and consolidation of morphotectonics, a Structural Geomorphology began to take shape and gain meaning in the works of Saadi (1998), Peulvast and Claudino Sales (2000, 2003, 2006, 2007) and ultimately, in the world of neotectonics and its relationship with the relief (Bezerra et al. 2001, 2008).

According to Bezerra et al. (2008), studies about the geomorphological evolution of theBrazilian Northeast are based on a pediplanation model, with morphology in response to uniform uplift and concomitant development of erosion surfaces.

Such a conception is not confirmed if the criteria of analysis, in addition to being topographical, are also morphostratographic and morphotectonic. There is increasing evidence from morphotectonics that the geomorphological evolution of theBrazilian Northeast occurred in a much more complex manner then that proposed by the pediplanation model, which is quite limited in relation to recent conceptions regarding intraplate tectonics. This is the case because the pediplanation model does not incorporate rifting mechanisms and the history of basins, a limitation derived from the idea of stability in Brazilian territory. This model also does not incorporate data on post-rift reactivation, restricting itself to a model of uplift and planation describing the equatorial passive margin to the east of South America and Western Africa as successive terraced surfaces, developed from an uplift and subsequent erosion.

Peulvast and Claudino-Sales (2003), addressing the morphotectonic evolution of theBrazilian Northeast, questioned the model of successive post-Cretaceous uplifts as responsible for developing, until the Plio–Pleistocene, successively embedded planation surfaces. For these authors, the relief of the Borborema Province occurs around a central depression, the "Jaguaribe depression", corresponding partly to the Jurassic–Cretaceous Cariri–Potiguar rift zone, with its morphology characterized by segments of marginal escarpments, equivalent to the extremes of the shoulders of the aborted rift. The Brasiliano shear zones controlled the main features of differential erosion, like escarpments and valleys on fault lines. It seems clear that the renewed interest in the role of tectonic geomorphology is a direct consequence of the assimilation of the concepts of global tectonics, which no longer allow for the notion of the existence of portions of the lithosphere endowed with absolute crustal stability (Saadi 1998).

Given this situation, recent studies have proven that intraplate seismic activity demonstrates Quaternary tectonic activity in the Northeast (Bezerra et al. 2007). Its relation with the relief is gradually being verified, some of these will be addressed below, and from a context that takes into account the consolidation of conditions of the Brazilian platform, its Cretaceous individualization, the neotectonics, and the impact of these factors on fluvial systems.

7.2 Tectonics and Fluvial Systems

The geometry of fluvial channels is the result of complex interactions between the type of load, the flow regime, topography, substrate, and tectonic activity (Schum et al. 2000). Such activity is regarded herein as neotectonic, given the fact that its effects are felt in Quaternary environments.

According to Saadi (1993), the neotectonic framework of the Brazilian platform features deformation in all its amplitude. These deformations, directly associated with the preexisting lines of weakness, are present at the height of the northeastern region of Brazil, characterized by a high number of active faults and seismicity.

The significant seismic activity in the Northeast region necessarily requires a geomorphological approach, a study about the effects of intraplate stresses and consequently, the effects of deformations in the crust, the sedimentary environments and in morphogenetic processes, which highlight the effects of periodic uplifts, continental flexure, and transcurrent faulting controlling the remarkable seismicity (Saadi et al. 2005).

In this context, an analysis of the Paleo stress fields and their possible influence on the organization of the drainage system becomes of paramount importance in updating the knowledge related to the geomorphological evolution of the Northeast, considering that river currents represent the main driving mechanisms of the landscape through dissection and alluvial action.

Therefore, the analysis proposed herein starts with the relationship between the geometry of fluvial channels and the direction of their flow with the configuration of the Precambrian basement and Cretaceous sedimentary basins. Empirically, there is evidence of the coherence between the direction of the preferential flow of the main rivers in the north of the Northeast and the direction of the structural trends.

The lower reaches of the river valleys (Cearense and Potiguar) are grabens generated by the reactivation of transcurrent shear zones (Saadi and Torquato 1994), and therefore, drain their runoff under strong structural control.

As to morphotectonic aspects, it stands out that much of the Quaternary sedimentation is confined in tectonic depressions. Examples of this relationship between sedimentation and tectonics are observed in the valleys of the Mossoro and Açu Rivers (RN). These valleys dissect Cenozoic (Barreiras) and Mesozoic (Apodi Group) sedimentary deposits sitting over asymmetric grabens (Potiguar Rift), whose origin is linked to variations in an equilibrium profile. These variations, in turn, are driven by climatic and eustatic changes, or by an uplift process of the Borborema Province active in the Cenozoic (de Sá et al. 1999).

The dissection in the coastal portions occurs on the rocks of the Barreiras Formation, classically interpreted as vast pre-coastal glacis. The morphostructural compartmentalization is mainly related to Cenozoic semi-grabens whose sedimentary fill is directly related to the denudation of the adjacent horst.

The dissection occurs in a differentiated manner according to the area analyzed, sometimes presenting sudden changes in the channel geometry. Thus, dissection and gradation may be evidence of tectonic activity, especially when accompanied by a change in the morphology of the channel (Schum et al. 2000).

However, the lower surfaces of the coastline have their dissection led by falls in the sea level in the upper Cenozoic (Bezerra et al. 2001). For these areas, the role of the neotectonic seems to be more related to the conditioning of the valleys and aureole erosion than incision and linear erosion processes.

In the Brazilian Northeast, NE major tectonic lineaments, developed since the late Brasiliano Cycle, are widely acknowledged. These lineaments control the meandering route of the rivers, where the preferred directions of flow are expressed as NE-SW and E-W lineaments, giving these a clear structural control.

In the study area, several points with the occurrence of faults affecting Cenozoic cover present significant correlation with patterns of lineaments and drainage anomalies. For example, drainage anomalies associated with changes in the morphology of the channel and tectonic control were identified in the Jaguaribe River Valley by Maia (1993), Maia (2005) and Gomes Neto (2007).

Indeed, it can be said that in the area analyzed the rivers drain their runoff undergoing a structural control of the drainage at a regional level. It is observed in Fig. 6.1 that generally the rivers preferably follow an NE-SW direction. The drainage tends to follow an E-W direction, perpendicular to the current coastline and parallelized to the structural planes, when the shear zones present this direction.

7.3   The Morphotectonic Evolution of the Brazilian Northeast

Two tectonic events are the main episodes of tectonic restructuring of the Borborema Province. The first occurred during the Brasiliano Orogenic Cycle that took place from the formation of the Gondwana megacontinent (Brito Neves et al. 1999, 2002). After the Brasiliano orogeny, the Borborema Province underwent a period of tectonic stabilization (Almeida et al. 2000) until the Mesozoic, when the lower Cretaceous tectonic (Waldenian Reactivation) of divergent character, separated South America and Africa and generated the Potiguar Rift (Matos 2000). This period was marked by the Brasiliano reactivation structures as well as major faults that formed the grabens and consequently all structural framework of the Potiguar Basin.

For the Cenozoic, stresses are related to the migration of the South American plate to the east and predominantly compressive intraplate stresses. These stresses were generated by the expansion of the ocean floor in the mid-Atlantic chain as well as in the Andean chain and are a compressive regime of E-W direction for the entire Northeast (Assumption 1992). See Fig. 7.1

Fig. 7.1 Seismicity in Northeast Brazil. Arrows indicate the current compression regimes. Seismicity according to Ferreira et al. (2008) and references cited therein (*Source* Maia and Bezerra 2010)

During the Cenozoic, there were events such as the reactivation of major fault systems, folding with large wavelength and axes oriented preferably in an N-S direction, resulting from E-W compressive stresses, which affected the Potiguar Basin in the Paleogene (Cremonini and Kender 1995).

High levels of terraces are commonly found in the main valleys. The origin of these terraces is related to dissections possibly induced by normal components associated with strike-slip movements that lower the erosive base levels and promote acute incisions in fluvial deposits. This finding is in agreement with the data on the effects of the Holocene global eustasia in the region, whose maximum transgression raised the sea level by 3 m relative to the current level (Bezerra et al. 2003).

The concept of normal movement associated with transcurrence was corroborated by Barreto et al. (2002), who found Pleistocene marine terraces (120 ka) uplifted from 10 to 12 m in the NE portion of the Potiguar Coast.

From this perspective, Fournier et al. (2006) described the coastal trays from the central part of the state of Paraíba as a piano keyboard, to identify the acute differences in the height of the trays, with enclosed valleys showing a gap in relation to the thalweg of over 100 m. For these authors, the post-Cretaceous reactivation generated distinct uplifts and tilting of geomorphic surfaces, where numerous faultings have streamlined indentations and guided the dissection.

Indeed, it was observed that from the Paleogene to the Quaternary, the bentonite deposits related to Barreiras indicate relationships with several pulses of uplifts of the crowning surfaces that limit their length inland (Saadi et al. 2005).

Bezerra et al. (2001, 2005) and Nogueira et al. (2006) corroborate this proposition to show that the faulting processes have affected Cenozoic deposits and verifying different thicknesses in this unit suggested syn-sedimentary faulting occasionally inducing liquefaction.

Regarding the occurrence of post-Pliocene tectonic activity, Bezerra et al. (2008) identified spasmodic colluviation processes associated with the reactivation of faults and the subsidence of grabens. According to luminescence data, such reactivation would have occurred as in two periods in the graben area between Cariatá and Paraiba, namely: 224 and 128 and 45 and 28 ka.

As an example of Neogene deformations associated with seismicity, Bezerra et al. (2005) have identified numerous liquefaction structures in Quaternary sediments associated with fluvial deposits of intertwined channels in Rio Grande do Norte and Ceará.

In the central portion of the Potiguar Basin, recent studies have revealed features indicative of reactivation and tectonic inversions of the main normal rift faults, ranging from the basement to the post-rift section, including affecting Neogene covers of the Barreiras Formation (Person Neto et al. 2008). In this sector, the geomorphological evolution is related to a possible inversion of the basin resulting from the effects of the (σ1) EW and NW-SE paleostressors operating in the Cenozoic. This inversion is characterized by a dome (Serra do Mel), prolonged in an NE-SW direction, limited to the N by the coastal strip, in the SW by the Poço Verde-Caraúbas shear fault zone, to the SE through the Açu River valley and to the NW, W, and SW by the valley of the Mossoró River.

These post-rift reactivations occurred in pulses and affected all the sedimentary sequences of the Potiguar Basin. Pessoa Neto et al. (2008) have recognized the existence of three pulses, the last of a post-Campanian age and therefore the most important from a geomorphological point of view, as it regionally affected the post-rift section. These reactivations have a direct effect on the characteristics of the drainage. In this respect, the valley of the Apodi-Mossoró River runs through active fault zones of the Potiguar Basin and is thus an important indicator of neotectonic through its morphology. In the central part of the Potiguar Basin, the topography of the valley of the Apodi-Mossoró River is characterized by two topographic highs (Serra de Mossoró and Serra do Mel, Fig. 6.4) bordering a topographic depression where its river plain is located.

These topographical highs are typical inversion features of basins resulting from the last post-Miocene stress field. In this aspect, the reactivation of the NE-SW fault systems in the direction of maximum horizontal stress σ1 NW-SE, resulted in the deformation of the post-rift sequence of the Potiguar Basin. This deformation is expressed in the relief in the form of structural highs (Serra de Mossoró and the Serra do Mel) in the central part of this basin. Among these structural highs, the formation of a depression in an NE-SW orientation enabled the expansion of the fluvial-marine system toward the interior of the continent, forming a fluvial-marine plain 25 km long by 8 km wide, thus creating the largest of the valleys that dissect the sedimentary Potiguar Basin. Thus, the dissection promoted by the general lowering of the base level in the last glacial peak reached the Cretaceous basement represented by the post-rift section of the Potiguar Basin. The depression formed from that dissection allowed the advance of the fluvial-marine system to the area of the lower course of the Apodi-Mossoró River. In this sector, the 0 m elevation enters about 30 km into the continent as shown in Fig. 1.4. This characteristic made it possible for this area to develop the most important Sea Salt production zone in Brazil, accounting for 50% of national production (Costa 2008).

These characteristics result from the control of the flow exercised by the Apodi Dam, located 110 km upstream of the river mouth and the escarpment of the top of post-rift section on the edge of the sedimentary Potiguar Basin that reduces the fluvial competence of the channel in the lower course and prevents progradation of the river system toward the shoreline. In this way, the fluvial deposits of the valley of the Apodi-Mossoró River, when not absent, are thin, as outcrops of Cretaceous basement in the area of the floodplain are common.

These thin alluviums make the Apodi-Mossoró River a sensitive indicator of the characteristics of the substrate.

So, the main channel expresses the current structural framework in its forms.

During the Cenozoic, there were events like the reactivation of major fault systems (Carnaubais and Afonso Bezerra), folding with very long wavelength and axes oriented preferentially in the NE-SW direction, resulting from NW-SE compressive efforts affecting the sedimentary Potiguar Basin in the Paleogene (Cremonini and Karner 1995).

In the central part of the sedimentary Potiguar Basin, an area corresponding to the lower course of the Apodi-Mossoró River, two fault systems were identified by Bezerra and Souza (2005). These fault systems (Afonso Bezerra and Poço Verde-Caraúbas, Fig. 4) of an NW-SW direction result from the latest post-Miocene stress fields. These fault systems sometimes affect neogenic units including the Barriers Formation and condition parallel-type drainage in an NE-SW direction for main channels (third and fourth order) and NW-SE for tributaries (first and second order).

Several drainage elbows interrupt the NE-SW parallelism of the main channel, forming small NW-SE segments and are important evidence of Quaternary reactivation. These faults document a stress field related to an EW compression and N-S extension compatible with a dextral strike-slip regime.

In the study area, several points with the occurrence of faults affecting Cenozoic covers have a significant correlation with patterns of lineaments and drainage anomalies. The anomalies detected are characterized by abrupt changes in the direction of the main fluvial course, the most frequent being: NE-SW to SE-NW.

The drainage elbows shown in Fig. 6.7 alter the preferred direction of the main channel (NE-SW) and are related to lineaments recognized from small incised valleys aligned along an NW-SE direction. However, not all elbows display a relation to lineaments recognized in aerial photos or Landsat TM 7 images, treated with

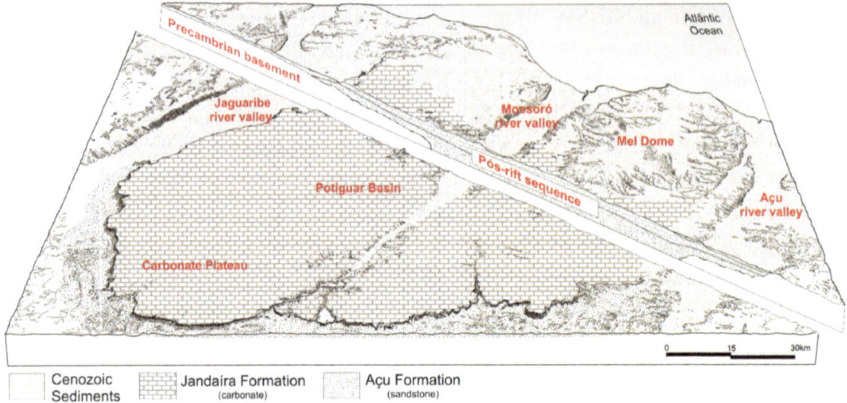

Cenozoic   Jandaíra Formation   Açu Formation
Sediments   (carbonate)   (sandstone)

Fig. 7.2 Relations between the sinuosity of the channel and fault zones in the lower course of the Apodi-Mossoró river

directional filters. These lineaments, imprinted on the topography, are related to the regional fault systems. The NW lineaments are the more expressive Afonso Bezerra and Poço Verde-Caraúbas (Fault Systems) and are marked on the relief in the form of incised valleys when they occur on the sandstones and conglomerates of the Barreiras Formation and ridges when they occur on the carbonate Jandaíra Formation. These lineaments influence the morphology of the Apodi-Mossoró River channel reducing its sinuosity values when crossing the active fault zones of the Potiguar Basin (Fig. 7.2).

The growing body of evidence about the Neogene–Quaternary tectonic activity in the Northeast requires the construction of an understanding concerning their effects on the development of the relief, as morphogenetic and morpho-evolutionary conditioning. The reactivations and their impact on the Neogene deposits suggest the need for an update of the classic interpretations and related delineation of a geomorphology of the Cenozoic, strongly mapped out by the quantification of the processes, an individualized interpretation, and definition of active geomorphic processes.

In sequence with intracontinental rifting accompanied by subsidence and oceanic opening, the formation of the passive margin and uplift of the Borborema Province, the morphotectonic evolution of the Brazilian Northeast in the Cenozoic was directly affected by events generating post-rift structures.

For the river systems of the Brazilian Northeast, it was found that the neotectonic is expressed in the form of control of the structural drainage, scaling of terraces and conditioning of valleys and deformations in Neogenic–Quaternary rocks and therefore the control of the morphological features.

Thus, the intraplate Cenozoic tectonics is itself an aim for geomorphology studies. The identification of morphologies related to post-rift events such as fault escarpments, structural surfaces, alignment of crests, structural valleys, deformations in neogenic rocks, liquefaction structures, and structural drainage control can consistently support the analysis of the geomorphological evolution of the northern North-east. This identification may also clarify the tenuous relationship between tectonics and the conditioning of drainage, with the development of the dissected edges and levels of the Borborema and their geomorphological evolution during the Cenozoic.

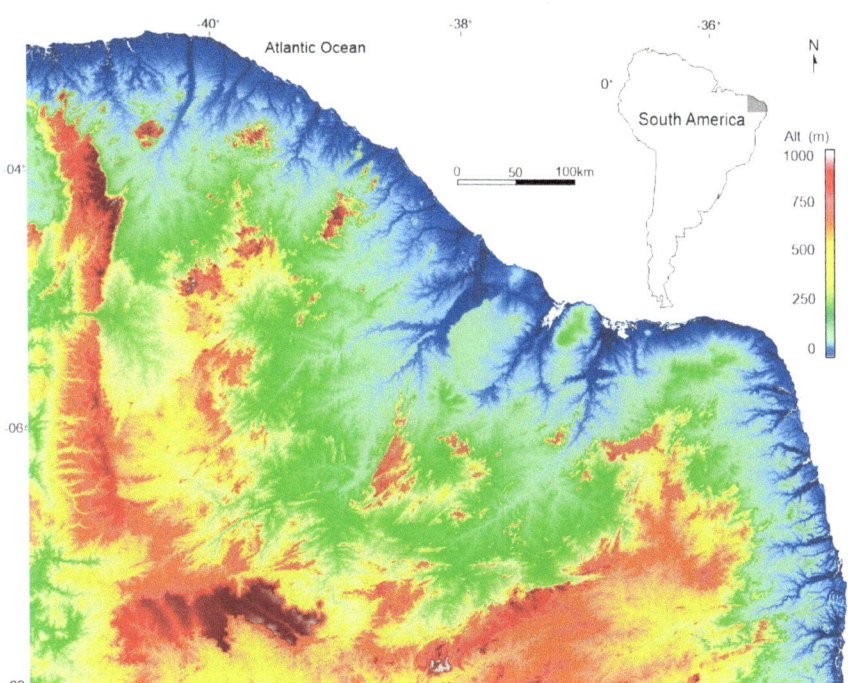

References

Ab Sáber AN (1960) Posição das superfícies aplainadas do Planalto Brasileiro. Notícia Geomor-fológica, SP. 3(5):52–54

Ab Sáber AN (1969) Participação das superfícies aplainadas nas paisagens do Nordeste Brasileiro. IGEOG-USP, Bol Geomorfologia, SP, n 19, 38 p

Almeida FFM, Brito Neves BB, Carneiro CDR (2000) The origin and evolution of the South American Platform. Earth Sci Rev 50:77–111

Andrade GO, Lins R (1965) Introdução à morfoclimatologia do Nordeste do Brasil. Arquivos do Instituto de Ciências da Terra, Recife 3–4:11–28

Barreto AMF, Bezerra FHR, Suguio K, Tatumi SH, Yee M, Paiva R, Munita CS (2002) Late Pleis-tocene marine terrace deposits in northeastern Brazil: sea-level changes and tectonic implications. Palaeogeogr Palaeoclimatol Palaeoecol 179:57–69

Bezerra FHR, Vita-Finzi C (2000) How active is a passive margin? Paleoseism Northeastern Brasil. Geol 28:591–594

Bezerra FHR, Amaro VE, Vita-Finzi C, Saadi A (2001) Pliocene-quaternary fault control of sedi-mentation and coastal plain morphology in NE Brazil. J South Am Earth Sci 14:61–75

Bezerra FH, Barreto AMF, Suguio K (2003) Holocene sea-level history on the Rio Grande do Norte State coast. Brazil Marine Geol 196(1–2):15

Bezerra FHR, Fonseca VP, Vitafinzi C, Lima Filho FP, Saadi A (2005) Liquefaction-induced structures in Quaternary alluvial gravels and gravels sediments, NE Brazil. Eng Geol. In: Obermeier SF (ed.), Paleoliquefaction and appraisal of seismic hazards, vol 76:191–208

Bezerra FHR, Takeya MK, Sousa MO, Nascimento AF (2007) Coseismic reactivation of the Samambaia fault, Brazil. Tectonophysics 430:27–39

Bezerra FHR, Neves BBB, Correa ACB, Barreto AMF, Suguio K (2008) Late pleistocene tectonic-geomorphological development within a passive margin—the Cariatá trough, northeastern Brazil. Geomorphology 01:555–582

Bigarella JJ (1994) Estrutura e Origem das Paisagens Tropicais, vol 1. Florianópolis: Ed. UFSC

Bigarella JJ (2003) Estrutura e Origem das Paisagens Tropicais, vol 3. Florianópolis: Ed. UFSC

Brito Neves BB (1999) América do Sul: quatro fusões, quatro fissões e o processo acrecionário andino. Bahia. VII Simpósio Nacional de Estudos Tectônicos, SBG, pp 11–13

Brito Neves BBB, Neto MCP (2002) Ciclo Brasiliano: discussão prefacial. Anais do XLI Congresso Brasileiro de Geologia, João Pessoa, Paraíba

Cremonini OA, Kraner GD (1995) Reativação mesozóica da Bacia Potiguar. In: Simpósio De Geologia Do Nordeste, 6. Natal. Anais do VI Simpósio de Geologia do Nordeste, Natal, pp 181–184

Davis WM (1899) O Ciclo Geográfico. In: Geomorfologia – seleção de textos. Vol.1 AGB USP, republicado em 1991. N 19

de Sá EJ, Matos RM, Neto JM, Saadi A, Neto OP (1999) Epirogenia cenozóica na Província Borborema: síntese e discussão sobre os modelos de deformação associados. VII Simpósio Nacional de Estudos Tectônicos, Bahia. pp 58–61

Demangeot J (1960) Essair sur le relief du Nordest Brésilien. Ann. de Geographie, Paris 69(372):157–176

Dresch J (1957) Les problèmes géomorphologiques Du Nord-Est Brésilien. Bull. Ass. Géograp. Français 263(264):48–59

Ferreira JM, Bezerra FHR, Sousa MOL, do Nascimento AF, Sá JM, França GS (2008) The role of Precambrian mylonitic belts and present-day stress field in the coseismic reactivation of the Pernambuco lineament, Brazil. Tectonophysics 456:111–126

Gomes Neto AO (2007) Neotectônica no Baixo Vale do Rio Jaguaribe, Tese de doutorado, Universidade Estadual Paulista Júlio de Mesquita Filho, UNESP, Brasil

King LC (1956) A Geomorfologia do Brasil Oriental. Revista Brasileira de Geografia, Ano XVIII no 2

Lima CCU (2000) O Neotectonismo na Costa Sudeste e do Nordeste Brasileiro. Revista de Ciência e Tecnologia. 15:91–102

Mabesoone JM, Castro C (1975) Desenvolvimento geomorfológico do Nordeste Brasileiro. Boletim do Núcleo Nordeste da Sociedade Brasileira de Geologia. 3:3–5

Maia LP (1993) Controle Tectônico e evolução Geológica/Sedimentar da região da desembocadura do Rio Jaguaribe. Ceará, Dissertação de Mestrado, Departamento de Geologia, UFPE, Recife

Maia RP (2005) Planície Fluvial do Rio Jaguaribe: Evolução Geomorfológica. Ocupação e Análise Ambiental, Dissertação de Mestrado Geografia Física UFC Fortaleza – CE

Maia RP, Bezerra FH, Sales VC (2010) Geomorfologia do Nordeste: Concepções clássicas e atuais acerca das superfícies de aplainamento. Revista de Geografia (Recife), 27:6–19

Matos RMD (2000) Tectonic evolution of the equatorial South Atlantic. AGU Geophys Monogr. In: Mohriak WU, Talwani M (eds). Atlantic rifts and continental margins, vol 115, pp 331–354

Nogueira FCC, Bezerra FHR, de Castro DL (2006) Deformação Rúptil em Depósitos da Formação Barreiras na Porção Leste da Bacia Potiguar. Revista do Instituto de Geociências-USP. Geol. USP Ser. Cient., São Paulo, vol 6, n 2, pp 51–61

Pessoa Neto OC, Lima C, Bezerra FHR (2008) Papel das Inversões Tectônicas na Formação de Estruturas na Bacia Potiguar. In: Congresso Brasileiro De Geologia 44, Curitiba. Anais do 44 Congresso Brasileiro de Geologia, Curitiba, CD-Rom

Peulvast JP, Claudino Sales V (2000) Dispositivos Morfo-Estruturais e Evolução Morfotectônica da Margem Passiva Transformante do Nordeste Brasileiro. III Simpósio Nacional de Geomorfologia, Campinas, SP

Peulvast JP, CLAUDINO SALES V (2003) Stepped surfaces and Paleo landforms in the Northern Brasilian Nordeste: constraints on models of morfotectonic evolution. Geomorphology 3:89–122

Peulvast JP, Claudino-Sales V, Bezerra FHR, Betard F (2006) Landforms and neotectonics in the Equatorial passive margin of Brazil. Geodin Acta 19:51–71

Peulvast JP, Claudino-Sales V, Betard F, Gunnel Y (2007) Low post-Cenomanian denudation depths across the Brazilian Northeast: implications for long-term landscape evolution at a passive continental margin. Global Planet Change 114:1–45

Saadi A (1993) Neotectônica da Plataforma Brasileira: Esboço de Intepretação preliminar. Geonomos, MG. 1(1):1–15

Saadi A (1998) Modelos morfogenéticos e tectônica global: reflexões conciliatórias. Geonomos n 6, UFMG, Belo Horizonte, pp 55–63

Saadi A, Torquato JR (1994) Contribuição à neotectônica do Estado do Ceará. Revista de Geologia, Fortaleza-CE. 5:5–38

Saadi A, Bezerra FHR, Costa FD, Igreja HLS, Franzinelli E (2005) Neotectônica da plataforma Brasileira. In: Quaternário do Brasil. Holos Editora. São Paulo

Schum SA, Dumont JF, Holbrook JM (2000) Actives tectonics and alluvial. Rivers Cambridge University. 290 p

Bibliography

Almeida FFM, Hasui Y, Neves BBB, Fuck RA (1977) Províncias Estruturais Brasileiras—Atas VIII Simp. Geol. Nordeste, Anais. Campina Grande, pp 363–391

Assumpção M, Schimmel M, Escalante C, Barbosa JR, Rocha M, Barros L (2004) Intraplate seismicity in SE Brazil: stress concentration in lithospheric thin spots. Geophys J Int 159:390–399

Bezerra FHR, Amaral RF, Silva FO, Souza MOL, Fonseca VP, Vieira MM, Moura-Lima EM, Aquino MR (2009) Mapeamento Geológico Regional: Folhas Macau: SB-24-X-D-II (1:100.000) CPRM, Natal, RN

Bezerra FHR, Srivastava N, Souza MOL (2010) Mapeamento Geológico Regional: Folha Mossoró: SB-24-X-D-I (1:100.000) CPRM, Natal, RN

Bigarella JJ, Andrade GO Contribution to study área of Brazilian. Quater Geol Assoc Am Paper 84:433–451

Bizzi LA, Schobbenhaus C, Vidotti RM, Gonçalves JH (2003) (Org.). Geologia, Tectônica e Recursos Minerais do Brasil–Texto, Mapas e SIG

Brito Neves BB, Fuck RA, Cordani UG, Thomaz Filho A (1984) Influence of basement structures on the evolution of the major sedimentary basins of Brazil: a case of tectonic heritage. J Geodyn 1:495–510

Can be used to infer the location and strength of Holocene paleo-earthquakes

CECAV (2008) Relatório demonstrativo da situação atual das cavidades naturais subterrâneas (Rio Grande do Norte). Cecav – Centro de Estudos, Proteção e Manejo de Cavernas. Instituto Chico Mendes de Conservação da Biodiversidade. Brasília

CPRM (2006) Mapa de Geologia do Rio Grande do Norte, Shapes e SIG. CD-Room CPRM (Serviço Geológico do Brasil)

Cruz JB (2008) Levantamento Espeleológico: Prospecção, identificação e caracterização de cavidades naturais subterrâneas no lajedo do Arapuá, Felipe Guerra/RN, tendo como suporte as geotecnologias. Monografia de Graduação, Geografia UFRN

De Castro DL, Medeiros WE, Jardim de Sá EF, Moreira JAM (1998) Gravimetric map of the north-eastern Brazil and adjacent continental margin: interpretation based on the isostatic hypothesis. Revista Brasileira de Geophysica 16:115–131

Lima MIC (2002) Análise da drenagem e seu significado Geológico-Geomorfológico, Cd Room, texto em formato digital, arquivo PDF, Belém

Lucena LRF (2005) Implicação da compartimentação estrutural no Aquífero Barreiras na área da Bacia do Rio Pirangi-RN. 151 f. Tese (Doutorado em Geologia)—Centro de Ciências da Terra, Universidade Federal do Paraná, Curitiba

Mabessone MJ (2002) História Geológica da Provincia Borborema (NE do Brasil) Revista de Geologia UFPE, 15

Maia RP, Bezerra FHR (2009) Neotectônica, Geomorfologia e Sistemas Fluviais: Uma naálise preliminar do contexto Nordestino. Submetido e aceito para Revista Brasileira de Geomorfologia

McCalpin JP (ed.) (1996) Paleoseismology. Academic Press, NY, 583 p

Miller KG, Kominz MA, Browining JV, Weight JD, Mountan GS, Katz ME, Minster JB, Jordan T (1978) Present day plate motions. J Geophys Res Atmos 83(B11). https://doi.org/10.1029/JB083iB11p05331

Monié P, Caby R, Arthaud M (1997) The neoproterozoic orogeny in northeast Brazil: 40Ar/39Ar ages and petrostructural data from Ceará. Precambr Res 81

Neotectonics in the Equatorial passive margin of Brazil. Geodinamica Acta (Paris)

Oliveira JP (2009) Caracterização da pluma de contaminação numa antiga lixeira com o método da resistividade elétrica. Dissertação de Mestrado – Universidade Nacional de Lisboa, Portugal

Passchier CW, Simpson C (1993) Porphyroclast systems ans kinematic indicators. J Estruct Geol

Rao NP, Tsukuda T, Kosuga M, Bhatia SC, Suresh G (2002) Deep lower crustal earthquakes in central India: inferences from analysis of regional broadband data of the 1997 May 21, Jabalpur earthquake. Geophys J Int 148(1), 1 January 2002, pp 132–138. https://doi.org/10.1046/j.0956-540x.2001.01584.x

Reis ÁFC, Bezerra FH, Ferreira JM, Nascimento AF, Lima CC (2013) Stress magnitude and orientation in the Potiguar Basin, Brazil: implications on faulting style and reactivation. JGR Solid Earth. https://doi.org/10.1002/2012JB009953

Semarh (2000) RN Secretaria de Meio Ambiente e Recursos Hídricos, RN. Plano de Estadual de Recursos Hídricos: Bacia do Apodi-Mossoró, Natal, RN

Sugarman PJ, Cramer BS, Christie-Blick N, Pekar SF (2005) The Phanerozoic record of global sea-level change. Science 310:1293–1298

Survey Circular 688 (12 p)

Takeya MK (1992) High precision studies of an intraplate earthquake sequence in northeast Brazil. PhD thesis, University of Edinburgh, Edinburgh

Van Schmus WR, de Brito Neves BB, Williams IS, Hackspacher PC, Fetter AH, Dantas EL, Babinski M (2003) The Seridó Group of NE Brazil, a late neoproterozoic pre- to syn-collisional basin in West Gondwana: insights from SHRIMP U-Pb detrital zircon ages and Sm-Nd crustal residence (T) ages. Precambrian Res 127:287–327